INTEGRAL PUBLISHING HOUSE

EARTH IS EDEN

An Integral Exploration of the Trans-Himalayan Teachings

Integral Religion and Spirituality Volume 3

By Jon Darrall-Rew & Dustin DiPerna
Foreword by Bruce Lyon

Integral Publishing House

Earth is Eden

Published by Integral Publishing House

First Edition

Cover Design by James Redenbaugh

Edited by Charlie Heslop and Jennifer Ciucci

Dustin Diperna author photo taken by David Vergne
Jon Darrall-Rew author photo taken by Beatrice Madach

BrightAlliance.org

ISBN 978-0-9862826-2-1

Printed and Bound in the United States

This book is dedicated to humanity,
To the single tradition that is our planetary awakening,
And the One Reality that is Infinite Awake Presence.

Jon Darrall-Rew is co-founder of Synergy Forum, an international initiative dedicated to the integration of spiritual awakening with social transformation. Between 2011-2014, Jon served as head of the international spiritual community, Shamballa School, which was founded by Bruce Lyon in 2001 to explore the third phase of the Trans-Himalayan teachings. Jon practices within the spiritual lineages of Dzogchen and the Trans-Himalayan teachings, and is a long-term student of Integral Theory. He teaches Tibetan Buddhism and Integral Spirituality in Berlin, and works with individuals as a life path guide. He is dedicated to supporting the emergence of a global spirituality that honours and integrates all the wisdom humanity has generated on what it means to be awake, integrated, and fully engaged with life. Jon lives in Berlin, Germany.

Dustin DiPerna is founder of Bright Alliance and Co-Founder of Synergy Forum. He is an author, group facilitator, and meditation instructor. For the past decade has been a student of Integral Theory and practices in the spiritual lineages of Mahamudra and Dzogchen.

He received a Bachelor of Science degree from Cornell University and a Master of Liberal Arts degree in Religion from Harvard University. Dustin remains committed to the development of integrally informed spiritual paths that make spiritual awakening accessible to all. He lives in California with his wife, Amanda, and daughter, Jaya.

Table of Contents

Part 6: The Journey into the Kosmos

Part 7: Sacred Earth

Part 8: Conclusion

Tables and Figures

Acknowledgements

This book was made possible by a generous gift of financial support from several individuals. We would like to fully honor and thank Olivia Hansen, Bruce Lyon, and Michael Valentine. We would like to additionally honor David Martin for the guidance he provided to us in our writing process. The belief that each of you has shown in us, and the unyielding love that you've offered, has changed our lives forever. We are eternally grateful.

Foreword
by Bruce Lyon

Two things about this book make me very happy and give me real hope for the future.

The first is the spirit in which it was written. It is the result of a communion of souls rather than a coming together of minds. Therefore the book has its own livingness that one can engage beyond the words and ideas. As such it is a demonstration on the human level of the relationships it speaks about in a kosmic framework. The spiritual traditions have long spoken about 'brotherhood' but I feel it is particularly the quality of 'sisterhood'—the emergent value sphere of the feminine in men and women—that adds real potency to this collaboration. There is an appreciation of what they call the 'We' space, the third entity that hovers over and births through their relating that calls to mind the sohbet of Rumi and Shams. These men have taken the time to truly and deeply 'meet' in spiritual friendship before making the attempt to bring together the content of their minds, traditions, and worldviews. From this grounding in Prior Unity, identified with the One Life, a deep synthesis naturally takes place and the result is a delightful dancing together of content and space, emptiness and form.

This is the second thing that excites me. Integral theory and the Trans-Himalayan tradition have been two of the most prominent influences on my journey of awakening. Seeing them so expertly and creatively bought together in this collaboration by Dustin and Jon is like watching two friends whom you both love fall in love with each other. Not only have they achieved a great work of integration, but that very integration has allowed them to open up new lines of understanding and inquiry. In particular, their treatment of the relationship between radical and evolutionary awakening is rich and enlightening. Esoteric teachings about other forms of awareness in the universe and speculation about the role that Earth may be playing in this kosmic[1] environment are grounded in a way that brings intelligent and coherent discussion to an area that desperately needs it.

Spiritual traditions contain great jewels of wisdom, tumble polished to a rare beauty along the river beds of their lineages. Integral Theory and modelling is a wonderful framework for resetting these jewels in a crown that truly represents a synthesis of the 'Way of Earth'. Once we have harvested the essential riches of the world's diverse approaches to our essential divinity then it will be time for a new development, and this current generation of young people will be at the forefront of it. Informed and enriched by the traditions, but integrating and transcending them, they will 'listen together' to the waterfall of always already Spirit that is pouring from the eternal into each moment of time and space.

There is little disagreement that the current time and space we are living in is critical. Conscious contact with the Infinite is one half of the equation—the other is the way in which that contact is understood and unpacked. Sometimes the difference between an enlightened individual or society and an insane one is not the experience they are having but the way that experience is translated into civilisation and culture. This is where radical awakening and evolutionary awakening can come together. Radical contact with the Ground of Being is always available but the evolution of a civilization must be ripe to receive the lightning and transform itself into its next evolutionary potential. This requires integration, synthesis and embodiment. An analogy from the realm of science is the Allen Telescope Array where thousands of satellite dishes are aligned and pointed towards the the kosmos in the search for extra-terrestrial intelligence. They need to be coordinated in such a way that they can receive signals together, creating one giant telescope, and they need to have the right software that can understand and decipher any signals they receive. The analogy in consciousness is a global group of awakened beings representing the harvest of the world's wisdom: aligned, attuned, and running the most advanced consciousness software available. The revelation that is ahead of us will be kosmic and applied globally in a way that celebrates and supports the diversity of all life on Earth. It will not be individuals initiating new rivers but a global group representing Earth as a living being, downloading straight from the ocean together.

Earth is awakening in both the radical and evolutionary sense. We are both the universal Infinite Whole, Self-remembering and Self-

---

1. In this book we follow the pre-modern traditional approach to spelling kosmos with a 'k', rather than a 'c'. While the word spelled with a 'c' relates to the observable physical universe, when spelt with a 'k' it encompasses the entire realm of form, on all planes and realms, spiritual and material.

realising, and we are awakening as a part of the kosmos—as Earth, evolving in the time and space fields of a sentient universe. I believe that the skilful balancing of these dual expressions of our divinity along with the deep embodied integration of masculine and feminine approaches to awareness will allow us to take our place in a kosmic brother/sisterhood.

Each generation has its own adventure in the one great adventure. I am part of the boomer generation that used its signature narcissism to dream big and shine brightly, everyone a star in their own corner of the galaxy. We talked a great deal about cooperation and communion but it was goal and not ground. Jon and Dustin don't have communion as an ideal to reach but draw on it as an innate part of their make up. They instinctively 'get' the galaxy and the power of the field in which they are emerging and the universal Life that they are expressing. That too makes me very happy. As far as hope goes…well, I don't really need that, but I like feeling it. It all works out. It always does. It was intense and a close run thing but in the end, Emptiness and Form came together and……nothing really mattered. *Really* mattered.

Bruce Lyon
Founder of Shamballa School & Shamballa Tantra
Author of *Occult Cosmology, Working with The Will, Group Initiation, The Mercury Transmissions, Agni Way of Fire*

A Note from the Co-Authors
Jon and Dustin

We first met in Giza, Egypt, on the rooftop of a guesthouse overlooking the Great Pyramid. We traveled to Egypt at the invitation of Bruce and Sharon Lyon, to whom we owe our gratitude for bringing us together.

The recognition of brotherhood between us was deep and immediate. Both of us were living and serving in the world from the same animating impulse of making a contribution to the emergence of an awakened, kosmocentric civilization and culture of Reality on Earth, and particularly its expression in the form of a universal spirituality. For Dustin, a spiritual mandate demanding the universalization of esoteric teachings was part of his being from as far back as he can remember. The most concrete recognition of his life purpose in this regard fully landed during the inaugural event of the Integral Spiritual Center hosted in Denver, Colorado, in 2005. It was there that a lightning bolt of vision provided him with an understanding that there would come a time when the most awakened, evolved and integrated leaders of the great wisdom traditions gathered together publicly to represent both Reality-realization and the embodied example of authentic spiritual depth to all humanity. This group, through their lived example, would then play a role on the global stage that offered guidance and stewardship for the future evolution of life on Earth. And Jon, from his earliest memories, had always carried encoded in the substance of his soul the instruction to hear what all the different cultures and traditions of the Earth understand of the divine. He had always envisioned that one day the diversity of that wisdom might be integrated and synthesized into a trans-lineage universal spirituality that honored all paths, optimized their gifts, and yet also transcended exclusive affiliation with any in naked Reality-identification.

Both of us, by the time we finally met each other, already represented strong points of light radiating out of two different but complementary galaxies of wisdom. Dustin comes from a background in Integral Theory, under the mentorship of Ken Wilber, as well training in Mahamudra and Dzogchen under the direction of Daniel P. Brown. Jon's path has been based in the Trans-Himalayan teachings, the wisdom stream that flows out of the work of Helena Blavatsky, Alice Bailey, and more recently Bruce Lyon, among others, with a community of awakened masters operating on the subtle planes. Jon has worked closely with Bruce Lyon, is a student of Integral Theory, and practices Dzogchen under the guidance of Khenchen Lama Rinpoche and Mahamudra and Dzogchen with Daniel P. Brown. Each of us had some working knowledge of the other's primary wisdom galaxies and felt deep resonance where they overlapped. We could sense immediately that there was work to be done together and much that would come through the synthesis of visions that we both held.

The first night we met, we sat on the roof of our guesthouse under the stars with other brothers and sisters from around the world. Isioma was with us from Nigeria; Max, Alessia, and many others from Italy; Maurizio, Consuelo, and Michele from Switzerland; Angeliki and Samantha from Greece; Olivia from the United States; Bruce and Sharon from New Zealand; Remco from the Netherlands; Aunkh and Nia from South Africa, as well as others from distant corners of the globe. There were twenty-three of us in total. As the night continued, the crowd on the roof grew thin, and we eventually found ourselves sitting with Bruce Lyon.

With a passion for truth ablaze, we began to talk about the ways that the Trans-Himalayan system in general, and its emergent 'third phase' teachings in particular, might interact with the Integral framework. As the stars shifted in the sky and the light show projected upon the pyramids began to fade, Bruce offered us a pearl of insight. He explained that the joining together of Integral Theory and the Trans-Himalayan teachings could be compared to a galaxy floating in the dark matter of space. Integral, he explained, is like dark matter. Dark matter is theorized in modern science to account for 84% of the total mass of the universe, and yet to be invisible to all instruments of observation. It is only detectable indirectly, owing to its gravitational effects upon visible matter. It provides the scaffolding, a dark backdrop of invisible structure that helps to frame anything that arises in its field. Although like dark matter it has mass, Integral as a system is virtually content-free. The Trans-Himalayan teachings, and all spiritual traditions for that matter, Bruce explained, are therefore like galaxies. They are the content that arises in space. Each tradition is a galaxy of light, dynamically accelerating in time through space, lighting up potentials for emergence, and orbiting a supermassive black hole of pure Reality-transmission at its core. The dark matter that is Integral provides a scaffolding to highlight, frame, and evolve the gifts of the various lineages of teaching and traditions. Something we found exciting about the coupling of the two together is that owing to the Trans-Himalayan teachings' trans-lineage universal perspective, they can provide content right the way through from egocentric to ethnocentric to worldcentric to kosmocentric altitudes. And additionally, owing to the sophistication of the maps articulated in its multidimensional kosmic orientation, the Trans-Himalayan teachings invite the Integral framework to apply its brilliance in as yet unexplored territories.

Bruce's metaphor landed deeply for both of us. And it was from that encounter, and those initial seeds of wisdom first planted on the rooftop in Giza, that this particular book and collaboration sprouted and took form. We spent the next week together, trekking through the deserts of Egypt, allowing the recognition of the work to which our souls had already committed themselves to arise by natural resonance, and envisioning what a potential book might look like if we were to write together.

As our collaboration began to take form, we soon realized how we on Earth are no longer imprisoned by apparent separation of time and space. Aspatial connection has become the new norm. Today's inter-connected web of technological utilities and lightning speed internet connections allowed the two of us to beam into each other's life-spheres several times a week through video conferencing. Jon wrote from the mountains of north Greece and then the city of Bristol in the UK, Dustin from the heart of the redwoods in Northern California in the United States; two points of light on opposite sides of the world collaborating on the exposition of a particular constellation as it arises out of dark matter. Periodically we would also meet in the Alps of Switzerland, where other threads of our shared work would take us.

After our first three months of writing we began to notice a fascinating process unfolding as our minds wrapped and morphed together. We began to realize that our working "We", the unique emergent inter-subjective field that was born and enacted through our collaboration, began to deepen, expand, and intensify. Specifically, we noticed that our working "We", which began quite naturally as a clear recognition of already established unity on the level of soul, continued to unfold the latent potentialities on the levels of inter-subjective mind, emotion and vision. We began literally dreaming together, working through details as we slept that made perfect sense when we would both meet again in the waking state. It was all quite remarkable.

Two points of insight concerning the two streams we represent continued to reveal themselves and coalesce more and more fully at the early stages of writing. First, Bruce's suggestion, and a point Ken Wilber himself has made on several occasions, proved true: Integral is best thought of as a content-free framework for organizing all domains of human wisdom into a coherent, universally accessible form. As such, Integral Theory has the unique capacity to honor interior and exterior, individual and collective perspectives, as well as every developmental level of our human potential.

The second relates to the source of the Trans-Himalayan teachings. These teachings have been given by liberated masters who have awakened and evolved into trans-lineage spaces of Reality-communion and multidimensional kosmic exploration beyond exclusive identification with any religious tradition. As such, we saw that they offer a unique body of content that helps to flesh out the Integral framework: one that is beautifully aligned with the universal orientation of Integral Theory itself.

Bringing the Integral and Trans-Himalayan streams together allowed us to see some of the incredible gifts they can offer each other, and the world. Integral Theory provides a revolutionary framework that can allow integration across perspectives and discernment between developmental levels. As such, it supports humanity's awakening and collective evolution to transcend and include ego-, ethno-, and world-centric perspectives into a kosmocentric *experience* that begins to truly allow a unified humanity to become an awakened participant in the kosmos. The Trans-Himalayan teachings flow from a community of beings who have already done that. As such, we quickly noticed that when the gifts of these two streams are coupled with each other, a compelling contribution to planetary awakening and evolution—and especially the emergence of a genuinely universal spirituality—can be made.

With these points in our minds and hearts as the writing unfolded, Dustin brought forth various integrally-informed framing aspects to the endeavor, while Jon brought the insights and wisdom of the Trans-Himalayan teachings. The alchemical synthesis of these perspectives was enacted with full consciousness that, over time, the kosmic grooves and evolutionary habits we found ourselves laying down would be further contributed to by our brothers and sisters around the globe. We envisioned that as time unfolded, and through the use of such a framework as Integral Theory, they too would offer the fruits of their own paths into a dynamic space of sharing, integration and synthesis. It is from here that an awakened kosmocentric culture and civilization on Earth, and a universal spirituality—what we have come to call the Great Human Tradition—can emerge.

Just as this is in essence a book on Integral Trans-Himalayan spirituality, the purpose of this series of books on Integral Religion and Spirituality is to offer a platform for other authors to write books highlighting the teachings of other traditions using the integral meta-framework as a common denominator. All of this will result, we hope, in the beginning formulations of a canon of truly universal spiritual teachings for all of Earth.

And so, it is from this platform of communion that we offer this piece to the reader.

Jon and Dustin
Winter Solstice 2012

Welcome
Jon and Dustin

Realities transmit varying levels of Truth depending on the person or persons bringing them forth into existence. The vision in this book contains insights readily accessible only to those souls for whom a dramatically new and updated expression of ancient paths relating to Earth and the kosmos is the call of their hearts. There will be some who pick this book up and don't make it past the first few pages. There will be others who push on, yet still find that the book's contents make little sense. For both of these groups of individuals this message remains locked; self-secret until the timing may be right and ripe in their lives to explore further. Others, sensing the invitation and having the capacity to go deeper immediately, may decide to sit with this book, unpacking the truths it contains. For these souls the call of curiosity may arise to meditate upon the various ideas outlined until greater clarity is reached. For others still, this book will serve as a much longed-for travel companion; a map outlining a pathless path through the glorious bounty of Reality, the Earth, and the kosmos. From this perspective, we as authors are like scouts, passing along the truths of the always-already established Unity our souls have been called to live; the wisdom, knowledge and empowerment we have gained through our own journeys; and sharing the insights we have been honored to receive from those who walk further ahead of us. Finally, there will be others who read this book who hold such a degree of wisdom and mastery that any errors in theory and ideas presented here will be immediately obvious. To these souls we are humbly open to your correction and feedback. Whichever category you find yourself in, we recognize and honor that you are here with us. In the fellowship of the One Earth Lineage of Reality in the kosmos, we welcome you. May this book benefit your life.

This book holds two primary intentions. In service of the all-encompassing, ever-victorious and enthroned Absolute Reality, the primary intention behind this book is to contribute to the empowered revelation of the sacred Purpose of the Earth in the kosmos. This naturally calls forth the emergence of an awakened, kosmocentric culture and civilization of Reality on Earth.

Secondly, we recognize the fact that in today's world the Revelation of Reality has been imprisoned by ethnocentric and often pathological expressions of religion around the globe. This ethnocentric imprisonment, and the tribalism that results, is one of the single greatest contributing factors to world conflict and suffering. As such, this book is a contribution towards the emergence of a universal spirituality on Earth in which all transmissions of the sacred are able to flourish beyond any ethnocentric boundaries.

We envision such a universal spirituality not as a homogenized mass in which the differentiations between points of wisdom in the spiritual traditions are lost. Rather, we see it as one that can honor and even optimize those points of differentiation from a collective space that recognizes that different cultures have pioneered different forms of expertise over the ages when it comes to awakening, transformation, and the exploration of reality.

We feel that the marriage of Integral Theory and the Trans-Himalayan teachings embodied in this book is deeply generative and creative. It results in a model of understanding Reality, the kosmos, the Earth, and transformation and development as a whole, which has great significance for the way we understand our planetary potential. Additionally, we feel that it offers precisely the capacity to both honor and integrate the wisdom cultivated in the various traditions, as well as to synthesize it into a form that gives some first glimpses of the what the trans-lineage universal spirituality that is to come might look like.

In terms of our writing process with this book, we naturally found ourselves deeply aligned when it came to the vast majority of its content. And yet, in order to honor both individuality and collaboration in the writing process, we found that explicitly listing in the contents section whether the chapters were written by Jon, or Dustin, or both of us, was the best approach to take. We found that engaging the writing in this way allows for the full expression of individual perspectives without them needing to be tailored to ensure consensus, and for our collaboration to be full without needing to curtail individual expression.

One aspect of what you will find in this book is an advanced map of relative reality. This map includes multiple potentials in your own spiritual development. It includes various streams of unfolding within your own being. It provides a look into existence from the perspective of multiple kosmic planes of existence—all engaged from radical awakening to Absolute Reality. Many of the ideas articulated in this book are ageless. The masters as well as the esoteric and mystery schools of Earth have long held knowledge and insight into these domains. Something we have attempted to bring, however, is the holistic synthesis and articulation of these ideas into a universal form offered in a language that is readily accessible to an awakening humanity thriving at the edge of a digital age.

Novelty of this kind means that, in large part, this book is a massive project of translation, an upgrade of sorts wherein we offer post-postmodern maps for a dedicated group of spiritual practitioners on the planet. Such beings are those who are ready to explore what a radically awakened, kosmocentric culture and civilization of Reality, expressed through a universal spirituality, might look like at this particular moment in history. As such, this book is not meant to be a substitute for direct realization. The deconstruction of relative reality to its Absolute Essence (as we articulate in our descriptions of vector 1– radical awakening—below), is and can only be one of direct experience. Maps, outlines, and vectors like the ones articulated here serve only as conceptual distractions if not seen through in every moment as expressions of the divine. May all of you reading this turn your mind upon itself to recognize the boundless Awake Totality of Reality that you truly are. You are the source, substance, and manifestation of everything Good. Beyond thought, personality, time, and individual consciousness, you are first and foremost the Monarch of Awakened Love. May you fully come to know your ever-present divinity.

The degree to which we are each influenced and molded by our respective lineage streams cannot be ignored. We realize that while the stages of radical awakening within different traditions may show incredible consistency, the cosmologies of the lineages within which that awakening occurs will color how it is understood and conceptualized by the realizer. Principally, as pointed to previously, in this text you will find a synthesis of Integral Theory and the Trans-Himalayan teachings. However, let this not limit the delivery or the breadth of wisdom streams that are integrated. Both authors have found themselves influenced by multiple other currents (Dzogchen, Christian, Jewish, Islamic, Ancient Egyptian, African, Shamanistic, Hindu, Taoist, Zen, South American, and teachings rooted in no tradition, among others). Although emphasis is certainly greatest upon the teachings of the Trans-Himalayan system, each of these streams finds its way into these pages as well.

It should be clear from the start that we do not see ourselves as defining the end result of this sort of endeavor, but rather as representatives of two streams simply instigating an initial conversation. We understand clearly that no two individuals are going to be able to fully articulate the type and scale of emergence we are here envisioning. Rather, we see that it will require the integration and synthesis of all the best and most vital wisdom that has gone before us into new, universalized holons by Earth's brothers and sisters coming from all lineages and backgrounds. Indeed, through the use of the Integral framework, we see this book as the first attempt to integrate, synthesize, and translate the essential points of the Trans-Himalayan teaching into a globally accessible form that can contribute to a universal spirituality.

Therefore, as two of the first suns of different galaxies to begin dancing in each other's stellar fields as their respective galaxies collide and their central cores fuse, we humbly offer what we can to this process. We are well aware that the only true way that Earth culture and civilization can begin to express its planetary Purpose is if representatives of every sphere of wisdom and knowledge—from the spiritual to the scientific to the economic, the ecological, the technological, and the educational—can stand together as brothers and sisters of humanity. Then, using a meta-framework like Integral Theory (or whichever meta-framework comes along that proves most efficient), each leader can speak as an insider of their respective domain. By doing so, they can coordinate and organize the gifts they bring according to deep universal categories.

In this way, we hope that future volumes of this Great Human Tradition series will continue to receive teachings from those members who represent true insiders and who can correct for any biases present here that the authors' blind-spots do not allow them to see. We realize with deepest gratitude and joy that our ability to begin to bring together the kind of synthesis of wisdom outlined in these pages is an expression of the profound streams of wisdom we have been so privileged to encounter in our lives. And we recognize that the very opportunity to co-create the future we each know in our hearts arises from the friendships we have been so delighted to forge with others of so many different global traditions and lineages. In this, we have come to understand that scholarship and study alone can only take trans-lineage dialogue so far. It is not until those of all traditions, nations, fields, and disciplines have found each other in planetary fellowship that the trans-lineage communion necessary to begin to reveal the one lineage of Earth in the kosmos can be birthed.

May this be the beginning of a full, dynamic and rich dialogue of true trans-lineage synergy. May we each come together as stars to form constellations of light around the Earth. And may we each stand fully awake, and present to all of our brilliancy as we form a true galaxy of love, wisdom and power together, as we circle the omnipresent core of Awake Light that is the living Base of all. May the true light that we are as a human species be birthed as it has been foreseen.

PART 1: INTRODUCTION

You and We—here together in spirit.
One Temple for all—for all, One God.
Manifold worlds dwell in the Abode of the Almighty,
And the Holy Spirit soars throughout.
The Renovation of the World will come—
the prophecies will be fulfilled.
People will arise and build a New Temple.

(Master Morya)[2]

---

2. Roerich, 1924. p. 9.

Chapter 1
Integral Theory
Dustin

Before Jon introduces the reader to the Trans-Himalayan tradition in the following chapter, I'll begin here by offering a basic frame for our exploration together. As I see it, we are all part of a single, on-going human lineage. For nearly all of our current historical narrative, this obvious fact has remained shrouded from recognition. Until recently on our developmental journey as a human species, nearly all of the models and concepts that we have used to view the world have created division. Tribalism, ego- and ethnocentricity have blinkered our vision and maintained walls of separation. Rather than recognizing our obvious unification as one great human family, we have remained divided into various streams according to particularized categorizations.

Now, we stand at the edge of evolution with a monumental opportunity in front of us. The time has come for us to welcome a new era, unified under a clear vision. This clear vision integrates rational discernment, optimistic prophecy and the divine pride of our human potential. Rationality allows us to fully honor the diversity and particularities of views that currently exist on our planet. There is little naivety in our quest. The rational lens helps to remind us that most human beings alive today still exist in a world-space where a unified view is not in their immediate experience. For all of these souls, may this vision of a single human lineage serve as a magnet of evolutionary attraction. May it serve as a beacon of future possibility.

For those already standing in witness of our single human lineage, may the optimistic prophecy that this book represents serve as permission to sink your toes even deeper into the rich soil of Eden. For it is from Eden that today's prophets tell the story of what is to become manifest. It is from Eden that Earth's prophets describe the world around them as it is already springing forth. The optimistic outlook offered here foresees a planetary culture that ushers all human beings into the fully unified human alliance, awakened to our place in the kosmos; a human alliance based on real developmental differences, real spiritual capacities, and above all else our shared potential for infinite possibility. Together, as we all stand firmly in Eden, we download a future of hope.

At the current time in human history, we have the sacred opportunity to begin articulating the contours of our shared potential. Through the grace of technology and the larger patterns of planetization we have access to all of the world's great wisdom traditions and mystery schools, as well as other forms of knowledge of astounding clarity from multiple domains. Many of the most secret and profound esoteric teachings from around the globe, once held closely protected within traditions (often as a necessity for their own preservation and survival), have now been made available as public information (Kabbalah and Dzogchen are two clear examples). Simultaneously, our understanding of the profound intricacies of the entire evolutionary process—astronomical, chemical, biological, psychological, and socio-cultural—continues to deepen. This means that now, not in spite of technology, but rather because of it, we can begin to assemble and unite the Great Human Tradition under a single and clear vision.

The recognition of our human unity can only occur if we understand and begin to synthesize the wisdom that has been discovered accross the globe concerning our human potential. What does it mean to be a human being? What do we know about the furthest reaches of our human evolution? What does it mean for an entire collective to awaken to Reality rather than just isolated individuals? Are we alone in this universe? What is possible for us as one planetary culture and civilization within this great magic of the kosmos? If answers to these questions are authentically sought and if a true synthesis of our world's great wisdom streams is desired, it is vital that we use a common platform of translation and organization; a framework and approach that can make sense of all of humanity's abundant diversity. For this, we turn to Integral Theory.

Integral Theory

Throughout this book, we use the broad meta-framework of Integral Theory to help organize the vast cosmology of the Trans-Himalayan teachings. Let me explain a bit about why meta-frameworks are so valuable for the sort of global synthesis that is beginning to emerge here on Earth.

Finding the Common Denominator

As we build towards a true integration of the wisdom streams of our planet, meta-frameworks, like Integral Theory, will become less of what appear to be a luxury and more and more of an obvious necessity. One way to see the benefit of a meta-framework is to understand it as serving the same function as *common denominators* as they are used in basic mathematics. Each of us was likely first introduced to the concept of a common denominator when we were learning to add, subtract, and compare fractional numbers. The use of common denominators in fractional mathematics and the use of meta-frameworks like Integral Theory allow us to compare seemingly different things according to a common context.

A brief example about how we use common denominators will help to refresh our minds as well as point to why meta-frameworks are so important. Let's say we want to add, subtract, or simply compare the fractions 1/2 and 2/3. The first step to working with two fractional numbers with different denominators (the denominator is the number on the bottom of the fraction) is to convert the fractions into new numbers of equal value so that both fractions share a common denominator. Once a common denominator is established the numerators (the numbers on top) can be added, subtracted, or compared based on the commonality of the denominator.

In the example above, the denominators are 2 and 3, respectively. When considering the conversion of the numerators and denominators we know that 2 and 3 share the lowest common denominator of 6. When the proper conversions are made, the first fraction 1/2 translates into 3/6 and the second fraction 2/3 translates into 4/6. As long as the fractions remain in the form of 1/2 and 2/3 (i.e., with different denominators) it is difficult to compare, contrast, add or subtract the two numbers. However, when each fraction is converted so that they share a common denominator of 6 they can each be compared according to their common basis. The fractions 3/6 and 4/6 can be added together, subtracted, compared and contrasted with ease.

In a similar way, all major arenas of human activity, and the knowledge they have generated, are like fractional numbers with different denominators. This is particularly true for the world's spiritual lineages. Each lineage arose out of different cultural contexts with varying elements of bias built into their systems of practice. Each tradition comes with its own basic assumptions and orientations towards reality. Each comes with dimensional awareness local to the field in which the impulse originated. Contexts and tradition-specific biasing factors are like the denominators of the fractions in the example above. As long as two lineages share different contexts and biasing factors they cannot be easily compared (i.e. their denominators are different). Most scholars would claim that comparing, contrasting or synthesizing different lineages without first establishing a common denominator (a meta-framework or shared context) is a major error (this was the mistake of modern scholars' attempts at establishing universals; see footnote for details).[3] When the deep structures outlined in Integral Theory are positioned as a common framework (or common denominator) across traditions, the various lineages can be compared, contrasted, and even synthesized where appropriate. This is crucial when we come to explore a universal spirituality that is able to honor, integrate, optimize, and synthesize the different areas of expertise cultivated by the different traditions rather than collapse them all into a homogenized mass.

Although this book draws from several spiritual and religious traditions at various points, and includes some content from our own speculations and experience, the main thrust of its insights come from the Trans-Himalayan teachings. Using Integral Theory as the base (common denominator) upon which the complexity of the Trans-Himalayan teachings can build will allow adherents of other lineages to compare, contrast, synthesize their own systems with the Trans-Himalayan world.

Although Integral Theory adds intrinsic value in and of itself to help deepen the cosmology outlined in this book, the use of Integral Theory as a common denominator becomes even more important for how this book might contribute to and be integrated into future

---

3. In modernity, centering around the end of the 19th century and the first half of the 20th century, spiritual lineages were compared and contrasted to each other in ways that lead to grand universalizing statements. Each lineage was treated as if it could be compared directly to others. This was like simply comparing the numerators of the fractions, while totally ignoring the denominators. Later, postmodern scholars where quick to point out that often the universal claims of modernists were made using the denominator (cultural context and biasing factors) of the researcher while totally ignoring the denominators (cultural contexts and biasing factors) of the "other". It wasn't until postmodern sensitivity to context, language, identity, and historical influence that researchers realized that they were comparing fractions with different denominators. This led some scholars to fall into a trap known as "relativism". A relativistic worldview, although positive in the sense that it correctly recognizes that denominators exist, erroneously concludes that because contexts are different, no comparisons can be made whatsoever. With the use of a meta-framework like Integral Theory, a theory that is primarily based on perspectives rather than perceptions, we can create a common denominator of shared meaning and understanding across different contexts. With a meta-framework in place, we can begin to synthesize universal systems that honor the particularities and sensitivities of postmodern inquiry, while stepping beyond relativism. We can actually, now, for the first time in history, compare and contrast the wisdom streams of Earth in a way that actually works.

projects oriented towards synthesizing spiritual wisdom streams around the globe so that a universal spirituality might be born. Because of this initial framing, this book will be ready for comparison and synthesis with future researchers who, using the Integral framework themselves as a common denominator, emphasize the gifts of other lineage streams. We imagine that future books in this series will set an Integral frame for Christianity, Islam, Buddhism, and Hinduism, among others, just as it has set the frame here to honor the Trans-Himalayan teachings. It is in this way that a truly universal spirituality based on common deep structures and basic perspectives can be built.

The following graphic points to the possibility of trans-lineage universal spirituality and how this book fits in to such a destiny.

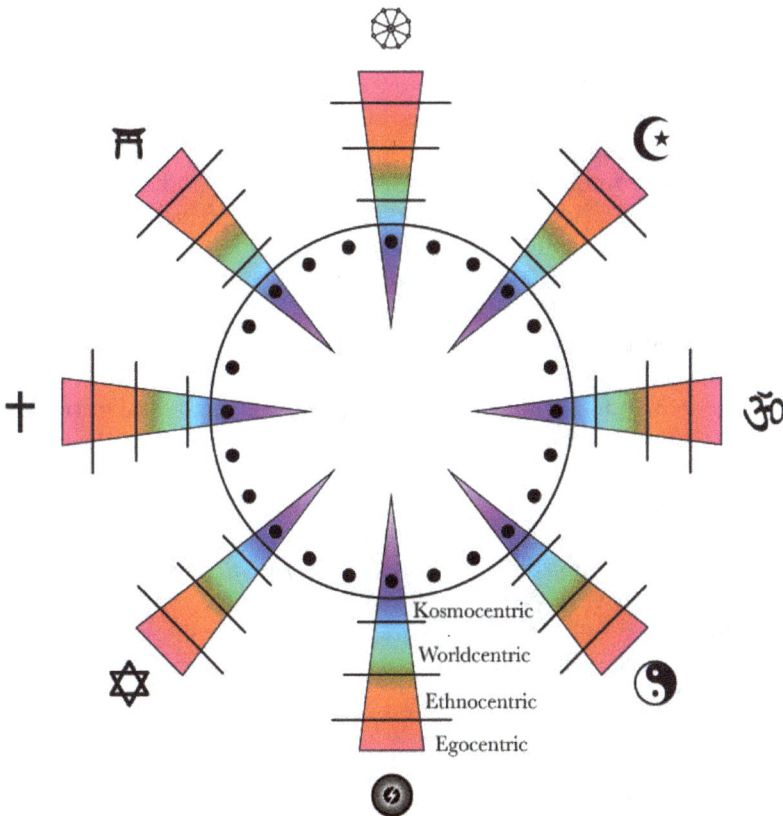

Figure 1. Graphic showing the pathway for each of the spiritual traditions from egocentric to ethnocentric to worlcentric to kosmocentric expressions. In the latter, they each offer

doorways into a universal spirituality. The points that do not stand within lineage streams represent those who have entered into a universal spirituality without particular connection to a tradition. The lightning symbol at the bottom represents the Trans-Himalayan pathway. The Shamballa School symbol is used to represent it as there is no single symbol for the whole tradition.

Each of the wisdom lineages of Earth, some of which are represented by the elongated triangles in the figure above, represents a stream in what we sometimes call the Great Human Tradition. This Great Human Tradition is the collective inheritance of wisdom from all traditions and pathways throughout the entire arc of human history. From the kosmocentric, integrative perspective, all of these various lineages are recognized as streams of a single river of wisdom. When ethnic, religious, and cultural barriers are seen through from the perspective of planetary synthesis, the recognition of the Great Human Tradition naturally brings forth a universal spirituality that optimizes the unique gifts of each lineage.

Far from some sort of postmodern syncretizing and homogenizing of the world's wisdom streams, this universal spirituality maintains the integrity of each stream, ensuring dual conveyor-belts of transformation remain intact (one moving from egocentric, to ethnocentric, to worldcentric orientations within each tradition along evolutionary lines, while another ensures the perpetual survival of transformative paths to radical awakening, from Gross to Subtle, to Causal, to Nondual realization). Simultaneously, a universal spirituality invites adherents of the various lineages to step nakedly into the trans-lineage center of a universal, kosmocentric trajectory. This book is an offering to that process.

With the context now set for how the offerings in this book can lead to the generation and revelation of a universal spirituality, the next few pages show how Ken Wilber's core elements of Integral Theory (quadrants, levels, lines, states, types, and shadow) can be used to help orient the reader to the basic cosmology of the Trans-Himalayan stream.

Applying the Integral Codex

Rather than going into extensive detail about each of these Integral elements here, we give only a brief frame and allow the text to explain the nuances of each element in the specific chapters dedicated to the task. To that end, we begin with a simple frame of how the basic elements of Integral Theory are operationalized throughout the text.

States

Generally, within the scope of Integral Theory, the category of states refers to Gross, Subtle, Causal phases by which Nondual Awakened Awareness comes into form. Reports of each these state/ phases are prevalent in almost all religious and spiritual traditions. Whereas the Quadrants, Levels, Lines and Types, discussed later in this chapter, relate to the changing and unfolding relative reality of the kosmos, we understand these states as relating to a spectrum that culminates in the recognition of Absolute Reality. In other words, these states relate to the way in which the Infinite Awake Presence of Reality arises as the entire kosmos. In the third phase Trans-Himalayan teachings, and in this book, these states have specific reference to the vector of radical awakening.

Radical awakening, as the Trans-Himalayan teachings explains it, is the path by which a relative self deconstructs relative reality to its Absolute Base of Nondual Awakened Awareness. This is the One Life or Boundless Immutable Principle of the Trans-Himalayan teaching. As radical awakening unfolds, Awareness literally witnesses the entire involutionary process unfold in each moment from Nondual Awakened Awareness, to Causal, to Subtle, to Gross manifestation. This means that Awareness has the capacity to rest as Absolute Nondual Reality while allowing all phases of its involution into form, which are continually unfolding moment to moment, to arise unimpeded. Specifically, this involves the initial distinction between subject-object that arises within boundless formlessness (the Causal state), the arising of the abundant and radiant light and energy that underlies all gross form to arise (the Subtle state), and then finally allowing all kosmic matter to arise (the Gross state), without confusing relative reality with Absolute Awareness. Succinctly put, in alignment with the traditions, we define radical awakening as awakening to and as the Awakened Awareness within which all states—Gross, Subtle and Causal—arise. As radical awakening is held open consciously during the waking state the latter gradually blends into dreaming (resulting in lucid dreaming) and deep dreamless sleeping (resulting in lucid sleeping). A fully radically awake being has conscious awareness of the involutionary process of Absolute Reality into relative form twenty-four hours a day. In this sense, the practitioner has gained full access to the ever-present universal base of Infinite Wakefulness.

According to both Integral Theory and wisdom traditions such as Vedanta, the first three of these states (Gross, Subtle, and Causal) can be seen as basic variations on the human sleep cycle. In a twenty-four hour period we move through states of waking (Gross), to dreaming (Subtle), to deep sleep (Causal). Integral Theory establishes the relevancy and universality of these states across multiple cultures and traditions with an important caveat: not all traditions and cultures have learned to develop conscious access and recognition of all of these states. Some traditions, for instance, stop conscious development at Gross level states, without going further to access Subtle, Causal and Nondual states. Integral Theory refers to such traditions as engaging "the path of shamans". The shamanic traditions, usually Gross realm orientated, guide their practitioners to open up access to states related to the natural world and the natural elements of the Gross realm. Some other traditions stop at Subtle level states without access to Causal and Nondual. Such paths guide their practitioners to open up access to Subtle levels of awareness and phenomena. These are commonly referred to as "the path of saints" owing to saints often having been recognized for their presence and penetration into subtle realms. Encounters with light, sound, visions, rapture, or contact with subtle masters or deity figures normally dominates their field of experience. Still other paths guide their practitioners into access to Causal level states but stop here without pointing to recognition of the Nondual root vantage point. These are referred to as "the path of sages". The experience of the sage takes experience to its pre-manifestation root—the Causal state. This is a void of absolute silence and quietude, prior to anything whatsoever arising. In Integral Theory, a Nondual path is often referred to as "the path of siddhas". The siddha traces experience to the Causal root and then allows all manifestation to arise (Subtle and Gross) as expressions of that root. The result is the direct experience of the unbounded single sphere of Reality-itself. The Trans-Himalayan teachings pay very careful attention to the dimension of all states, and the path of radical awakening it describes extends all the way through to full Nondual realization.

Quadrants

The entire text of this book is implicitly informed by Wilber's Four Quadrants. The Four Quadrants are formally introduced in Chapter 5. There, the quadrants are used in a traditional way to describe the fact that every sentient being has an interior and exterior, and both individual and collective dimensions of their being. This notion is presented throughout the text with reference to the dimensions of perspective used in language: I, WE, IT and ITS, or what we describe in terms of the four sub-lines of the evolutionary awakening vector of spiritual growth: identification, relationships/culture, bodymind, and plane access, respectively. As we shall see, the Trans-Himalayan teachings have an intuitive understanding of each of the Four Quadrants.

Levels

We describe levels (also referred to as structures, stages, altitude, and waves) throughout Part 4 on Evolutionary Awakening. One of the most profound points of the Trans-Himalayan teachings relates to its presentation of an infinite kosmic spectrum of planes of reality, extending from what are described as physical-etheric to emotional, mental, buddhic, atmic, monadic, and logoic planes, and beyond into the kosmos. We describe each of these levels/dimensions of reality in turn and describe how various orientations (1st, 2nd, 3rd person) influence the way in which we access and enact various elements of each stage. As we shall see, each of these levels of evolutionary unfolding can be correlated with a particular aspect of Integral Theory that Wilber refers to as altitude.[4]

Lines

In his book *Integral Psychology*, Wilber introduces three different groups of lines (specific areas of development). He lists (1) self-related lines, (2) cognitive lines, and (3) skills and talents. Chapters 10, 11, and 12, describe various levels of development along specific lines of the evolutionary awakening vector. These lines are best thought of as self-related lines. In those chapters, we consider an identification line (Chapter 10), which unfolds as a being's identification shifts into increasingly transcendent and inclusive levels of identity, from what the Trans-Himalayan teachings refer to as the personality, soul, and monadic bodyminds, and then further out into the kosmos. We then

---

4. In the most recent iterations of Integral Theory, and building upon work that Dustin released in Volume 1 and Volume 2 of this series, Wilber refers to four vectors of spiritual growth: structures, structure-stages, states, and state-stages. In our work in this book we make the following correlations: Structures = planes; structure-stages = levels of evolutionary identification; states = state/phases of manifestion; state-stages = vantage points.

consider a relationship line (Chapter 11), which unfolds as a being cultivates and opens up relationship with other beings residing on multiple planes of relative kosmic reality. We consider a plane access line (Chapter 12), which unfolds as beings gain access to the energies and information of those planes. In Chapter 13 we explore a group of lines that trace the unfoldment of stages of intelligence cultivation within and across the planes. As we shall see, each of these lines of development (whether self-related or cognitively-based) often unfold unevenly. A being can be highly developed in one particular line, while less developed in others.

Types

The Trans-Himalayan teachings use a complex typological system. Unlike more familiar psychologically-based systems of typology like Myers-Briggs or the Enneagram (both of which consider types at the level of the personality), the Trans-Himalayan cosmology uses a multi-dimensional system according to multiple levels of identification. It offers these distinctions using a cosmological system known as the *Seven Rays*. At lower-level planes of reality, the ray typology emphasizes the personality. At mid-level planes of reality the typology emphasizes the soul. At higher-level planes, the ray typology emphasizes what the tradition calls the monad—our indestructible essence.

Trans-Lineage Synthesis

It is our deep hope that the use of Integral Theory as a common denominator (here in this book and in others like it) will begin to allow the world's religious and spiritual systems to be compared to each other in a way that is sensitive to cultural context, biasing factors, and developmental levels. Using Integral Theory in this way, based on simple perspectives and deep structures will, we hope, provide the common underpinnings for a true universal spirituality to take form. It is our hope that such a spirituality will fully honor the particulars and the uniqueness of each tradition while simultaneously providing the frame for deeper synthesis and collaboration. May the Great Human Tradition, the lineage of One and All, take full form here on Earth.

Chapter 2
The Trans-Himalayan Teachings
Jon

In this book, you will find a body of content from the Trans-Himalayan teachings, explored through the perspectives of an Integral framework. Though there is necessarily a lineage-specific focus, it is our hope that the service this work may contribute to the world will have some significance across lineages and culture. Indeed, this is the intention that lies at the heart of both the Trans-Himalayan teachings and Integral Theory. Both, in their own unique ways, are orientated toward the awakening and evolution of all humanity and the revelation of the entire kosmos as intrinsically sacred. Because there are many readers who may not have come in contact with the Trans-Himalayan teachings until now, it feels appropriate to begin with a brief introduction to them so as to illuminate the context out of which the present work is emerging.

Over the course of their near 150-year history, the Trans-Himalayan teachings have played a significant role in awakening humanity to Absolute Reality; to the wonder of the living, sentient universe of which it is a part; to the birthing of a universal spirituality that is able to truly honor all paths; and to the unique Purpose that it is the role of the Earth to express in the kosmic context.

Fundamentally, they are a body of teachings that are understood to have flowed into humanity from a community of awakened and liberated sages, siddhas and bodhisattvas, dwelling principally in the Himalayas. These beings are understood to represent one branch of an ageless global lineage of wisdom and awakening on Earth. This lineage is one that is understood to have existed on Earth as long as humanity has existed here and even before, and to have preserved the wisdom concerning the True Nature of Reality, the multidimensional kosmos we are a part of, and the planetary Purpose of the Earth.

In the Trans-Himalayan teachings, this planetary lineage is known as Hierarchy, which this book will go on to explore extensively. The purpose of the Trans-Himalayan teachings is to support and empower the emergence of a radically awakened, kosmocentric culture and civilisation of Reality on Earth, and its expression through a universal spirituality.

While those beings whose teachings have formed the core of the Trans-Himalayan canon are understood to be based in the Himalayas, branches of this planetary lineage are described as existing, and to have existed, all over the world. They include not just the fastness of the Himalayas, but other places of power on the Earth such as those in Central and South America, Southern India, Ancient Egypt, China, Europe, Africa and Australia also. Those who have and do participate in this lineage are understood to be beings who have entered into liberation through the gates of all traditions and none in ages past, and who collectively form a planetary meta-sangha[5]. This meta-sangha is described as existing principally on the subtle planes of soul and spirit so as to empower and guide planetary awakening and evolution. That guidance extends through every evolutionary life sphere of planetary life, human and non-human, and includes every major arena of human activity—political, spiritual, educational, informational, economic, cultural, scientific, religious, and environmental.

The "Ageless Wisdom" that these beings represent is understood to have been the living seed at the esoteric heart of every global wisdom tradition. It is important to note that when we speak of this, we are speaking of an utterly different kind of lineage compared to the often mentally separative and ethnocentrically-oriented religious and spiritual traditions with which we may be more familiar. The planetary lineage we are speaking about here is one freely engaged by beings who have journeyed so deeply into the Mystery of Reality that they no longer operate through mind and mental forms of knowledge as we know them. Rather, their modes of perspective are based in the forms of wisdom-insight that flow naturally from direct contact and identification with Reality once the mental plane has been transcended. This necessarily entails shifts in polarisation into deeper layers of bodymind, community and plane than that through which the majority of humanity operates.

The content of the teaching found in the Trans-Himalayan transmission could be understood to have two identifiable sources. The first is the open release into the world of teachings that have likely been

---

5. Meta-sangha is a term coined by Terry Patten to refer to a collective community of practice and trust that includes many diverse sub-communities of spiritual engagement.

held secretly and shared only with the innermost initiates of the global wisdom traditions owing to them being of such an esoteric nature. This relates particularly to those traditions that have been geographically located in such remote and largely inaccessible parts of the world that their teachings have remained protected and intact for millennia. The existence of teachings that remain secret to those outside the initiated circles of a tradition (and about which those who *have* been initiated are forbidden to speak) is very real, even in our modern informational age in which so much of the Eastern wisdom has become easily accessible to the West. Many of the Vajrayana and Hindu esoteric teachings, for instance, are written in what is called "twilight language" (*sāṃdhyābhāṣā*). This is a polysemic system of language composed of symbols and images incomprehensible to those not initiated into their meaning. There are additionally a vast number of untranslated Tibetan and Chinese Buddhist texts, which are only known to a select few. It is quite possible that when and if these texts do begin to be translated, we will find that there is a great deal more wisdom and knowledge held in the esoteric cores of the wisdom traditions than many of us can currently imagine.

The second type of content comprised in the Trans-Himalayan teachings could be described as information not necessarily rooted in any human tradition as yet, esoteric or exoteric. Rather, the source of these insights are realisations and recognitions concerning the nature of Reality, the multidimensional kosmos, and the other communities of lives that are also awakening and evolving within it, accessed by a global community of masters. These insights are generated from altitudes of evolutionary development that are so advanced in comparison to our own that it is beyond common capacity to imagine, and that exist as the basis of their collective culture of shared wisdom and interrelation in planetary service.

The Kosmic Address of the Trans-Himalayan Teaching

All true spiritual teachings point to the existence of a sacred Reality that transcends, includes and permeates the time and space in which they arise. When coming to understand the unique gifts and significance of any body of spiritual teaching, however, it is important to consider the relative dimensions of perspective that they are coming through, as these will allow us to situate their contributions and limitations. This allows us to move beyond universal truth claims, and thus to be able to see the particular flavors of a teaching that might be able to be integrated and synthesized into a universal spirituality.

The Trans-Himalayan teachings have emerged at the point in our collective evolutionary journey when culture and civilization were first, through the rise of modernity, beginning to transition from ethnocentric into worldcentric spheres of sociocultural development. According to models of human collective development such as those advanced by such theorists as Jean Gebser and Jürgen Habermas, and then built upon by Don Beck and Ken Wilber, the transition from pre-modern to modern involved the differentiation of the value spheres of religion, science and ethics. This broke the stranglehold of ethnocentric religion on humanity's collective development, and the ideals of freedom, equality, and justice for all—regardless of race, creed or financial status—began to become the ideological basis of Western civilization.[6]

No longer were human beings necessarily united or separated by their belief in the same God and the domination of His Church. Now, science was able to progress freely through the gathering of a cross-cultural community of peers and the retraction of any part of nature or the kosmos as being "off limits" for investigation (thus allowing such advancements as the Copernican Revolution, for instance). With the rise of modernity, the ethnical injunctions according to which human beings determined "the good life" were no longer fenced in by the walls of ethnocentric alignment with one god to the exclusion of all others. This allowed certain major boundaries between human beings of different nations, classes, genders, and religious creeds to begin to fall, and we see this in such unfoldments as the abolition of slavery and women gaining the vote. By the end of the 19th Century the advancements of modernity, expressed in worldcentric, rationalist culture and industrial society, were beginning to enter the maturity they would fully reach in the next 100 years.

It was into this soil that the first seeds of the Trans-Himalayan teachings were sown, and I would argue that this is significant in relation to their capacity to contribute effectively to the emergence of a universal spirituality. It is without doubt that many of the founders of the great spiritual traditions in previous times (e.g. Gautama Buddha, Jesus, Lao Tzu, Padmasambhava, Shankara, etc.) were participants in the community of awakened planetary guides and stewards above mentioned as Hierarchy. Owing to the center of gravity of humanity's collective evolution in consciousness at the points in history when they gave their teachings being ethnocentric or lower, however (an example relevant to what we will go on to describe as the LL quadrant), and the

---

6. The degree to which Western civilisation has stayed true to these worldcentric principles is, of course, a worthy debate.

lack of the technological means for their teaching to be disseminated amongst the whole of humanity (an example relevant to what we will go on to describe as the LR quadrant), the possibility that it would be able to have an immediate impact globally was small.[7] Their teaching could only ever be expected, at least initially, to reach a particular ethnic population (the population of human beings residing in the Middle East in the time of Jesus, for example). We would argue that this would have been recognized by such awakened founders of the traditions, and they would thus have deliberately formulated and presented their teaching in a form that would make it most accessible to the particular ethnic population it would be given into.[8]

The Trans-Himalayan teachings, however, have been released at a time when the collective centers of gravity in human culture and civilization have provided the masters who transmitted them with very different opportunities. Specifically, they were released at a time when the collective cultural center of gravity was just entering the modern, rationalist era. Integral models of socio-cultural development understand this as the first point on the evolutionary spiral at which truly worldcentric perspectives begin to take form among collective cultures. In terms of the structure of society, they were released at a time when the level of collective technological advancement was such that it began to be able to support the flow of information globally. Both of these advancements would have been crucial to being able to make a significant contribution to the eventual emergence of a single culture and civilization of Reality on Earth, and its expression through a universal spirituality. According to their sources, the Trans-Himalayan teachings were released at the time they were for this very reason.

One aspect where the Trans-Himalayan teachings exemplify something rare on the global spiritual scene is in lieu of their source being a *community* of awakened masters, rather than just one single being. This is significant. It means that not only is the content of that teaching radically awake to the True Nature of Reality Itself, and coming through kosmocentric structures of consciousness, but that they are also not just the product of one individual's realization. Rather, they have emerged among a trans-lineage awakened collective culture

---

7. Note that there may be exceptions here. The argument can be made, for example, that Buddhism entered the world through the world-centric lens of the Buddha. However, the fact that world-centric systems were not available in the LR made the opportunities for global distribution of the Buddha's teaching significantly smaller than that which is possible today, where evolution in the LR has caught up.

8. When world-centric teachings were given, as they were in a number of instances, those who received them often immediately translated them down to ethnocentric interpretations.

(the Himalayan Branch of Hierarchy), where the realizations of each master have been cross-validated among the community prior to their dissemination. This perhaps embodies one of the first awakened and kosmocentrically evolved examples of Wilber's three strands of deep science serving as the basis for an entire line of teachings.[9]

Germane to this point and how we maturely respond to its nuances is the suggestion made by Ken Wilber that any trans-lineage spirituality that emerges would need to be *post-metaphysical*. That is, it would need to go beyond the pre-modern assumptions that reality has certain given levels and contours (as is entailed in the pre-modern teaching on the Great Chain of Being) that are the same for all beings. It would need to inculcate both the demand for evidence that came with modernity (as expressed through the rise of science), and the recognition of post-modernity that perspectives always arise within inter-subjective contexts that color and qualify their nature (as expressed through the humanities). According to an Integral understanding that honors both of these recognitions and takes them further, a post-metaphysical universal spirituality would understand that relative reality is not an ontological given that exists independently of its perceiver. Rather, it would see reality as enacted according to the developmental altitude (what level of development) and quadrant-focus (which quadrant is the focus) of a particular individual or collective, perspective.

This is what Wilber refers to as the "kosmic address" of any particular occurrence, such as the content of a lecture given in a university, the experience of laughter shared between friends, or the emergence of a

---

9. Ken Wilber has suggested that rather than spiritual experience being antithetical to rational empirical inquiry our understanding of it can actually be enriched by the scientific method. In his book, *The Marriage of Sense and Soul* (1998), Wilber builds upon William James's advocacy of a radical empiricism by proposing that the scientific method can just as easily be applied to investigate subjective experience as it can for objective matters. To unpack this, he presents the scientific method as being comprised of three essential steps that can be engaged to investigate any occurance, including spiritual experience. These are, a) perform an experiment; b) gain data from having performed that experiment; and c) check that data with a community of peers who have also done the same experiment. Wilber argues that despite modernist Western notions that designate all religion and spirituality as pre-rational and mythic with the scientific method embodying an evolutionary update to this, the post-rational esoteric traditions of mystical contemplation and meditation (e.g. Vajrayana, Vedanta, Zen, etc.) have always employed a form of the scientific method to investigate and consider the empirical reliability and validity of their realisations. This point is important as it excavates the designation of all religion and spirituality as pre-rational myth. While not denying that much, if not most of humanity's engagement with spirituality and religion is pre-rational and mythic, it allows recognition that there remain deeper forms of engagement that allow for not only rational but also post-rational approaches that include and trancend rationality in contemplative insight.

body of spiritual teachings such as the Trans-Himalayan. To find the kosmic address from which the Trans-Himalayan teachings have been transmitted, we need to look to both their source and receivers. That source is, as said, a community of liberated masters and bodhisattvas who form part of the Himalayan Branch of Hierarchy, particularly the Masters Djwhal Khul, Morya, Rakoczi and Hilarion. These are beings who are understood to be operating through the soul and monadic bodymind structures of our multidimensional kosmic identity, the details of which we will go on to explore in Chapter 10. Additionally, they form part of the collective culture of the Himalayan Branch of Hierarchy, and operate within its energetic fields. As such, it is important to note that the transmission given by the Trans-Himalayan masters will be expressive of their developmental level of attainment, both in terms of its extraordinary capacities and particular limitations. It will also be colored by the particular areas of unique wisdom and bias of the Himalayan Branch of Hierarchy and the information and energies contained within the collective energy fields of Hierarchy as a whole.

The collective culture of the Himalayan branch of Hierarchy here mentioned is just one Branch of the Earth Hierarchy, and as such will have its own biasing perspectives that might not be shared across the various other branches of Hierarchy, such as the African, South American, or Australasian Branches, for instance. This is an important point to remember as we proceed in our exploration of the Trans-Himalayan teachings, where particular perspectives on Absolute Reality, as well as the various relative layers of our multidimensional kosmic identity, the planes of existence and the communities of lives residing on them, are given. The perspectives entailed in this teaching should therefore be understood as those of a particular community of beings who, even though they have awakened and evolved to levels that the majority of humanity is unaware are even possible, still contain their own relativities.

In considering the kosmic address of those who received the Trans-Himalayan teachings, this involves taking a good look at such individuals as Helena Blavatsky, Alice A. Bailey, Helena Roerich, Lucille Cedercrans, and Bruce Lyon. If we look to the first four of these individuals—leaving Bruce Lyon for later owing to him living in a different cultural space and historical period from the first four—we can see that each of them was operating at the level of being an integrated personality with a clear opening to the soul level of their being. Additionally, these first four operated as part of the predominantly white European culture of humanity, with all the various sub-cultural spheres which that contains (e.g. strongly Christian influenced), in the historical period of the late nineteenth and early twentieth centuries. As such, they were functioning primarily through ethnocentric and rational-modernist mental structures (amber and orange for those familiar with Integral Theory). This means that although the content of the teachings were kosmocentric, there were and are to be found within them artifacts of language and social practice that surely contain traces of ethnocentricity and a modernist orientation that seeks to impose its worldcentric vision of the world cross-culturally. This is all to say that even if the content of the Trans-Himalayan teachings was transmitted by a group of radically awakened and kosmocentrically evolved masters, it was enacted and filtered through a particular set of individual, cultural and era-specific lenses as it came into the world via those who received it.

When it comes to Bruce Lyon and his reception of teaching from the same liberated Master that Alice Bailey also worked with (Djwhal Khul), we can see that now there are different biasing perspectives in play as a function of both his individual level of development and of the socio-cultural milieu within which he has been operating. In terms of Bruce's individual development, the privilege of close relationship and brotherhood that I have shared with him would have me suggest that this extends to his being deeply integrated personally, evolutionarily awake to the soul depth of his being, in clear contact with the monadic altitude, and with a stable opening to radical awakening. In terms of the mental structures predominantly present socio-culturally during which Bruce's work with the Master Djwhal Khul has unfolded, these have largely now unfolded into the post-modern cultural wave (symbolized with the color green in Integral Theory), with Bruce operating through integral structures of development (yellow, turquoise, indigo, and beyond). In accordance with the correctives post-modernity has brought to modernity, this has resulted in the teaching he has disseminated necessarily incorporating a much higher level of cross-cultural sensitivity, a deeper recognition of the inter-subjective relativity of perspective, and the need to allow others to speak their own truth rather than determining it for them. It has also generated some of the challenges inherent in the post-modern approach. For example, there has been some difficultly in being able to integrate the inherent worth of all beings with differences in the relative worth of their perspective as a function of the level of developmental depth they are coming from.

All of these reference points help us to situate, so as to begin to fully understand, the Trans-Himalayan teaching in its different expressions. All of them allow entry into understanding the unique types of wisdom that have been transmitted and received, as well as particular biases and relativities that it is important for us to note as present within them. They show us why it is so important with all teachings to acknowledge the difference between the enacting capacities across the entire AQAL matrix (quadrants, levels, lines, states, types) from which these teachings were being transmitted, and that into which they were being received. Otherwise, we might easily be tempted to throw the baby out with the bathwater. However, when this difference is recognized, we can clearly see when and if any ethnocentric, modernist, or post-modernist biases have come into play and, while honoring the individual in question, we can leave those perspectives at the door. This is a point relevant for how we interact with all teachings.

The Trans-Himalayan teachings offer a unique transmission of kosmocentrically framed Reality-wisdom that is of deep significance to humanity as a whole. As the above will hopefully have demonstrated, this uniqueness contains both extraordinary gifts as well as definite limitations. For the emergence of a universal spirituality to be fully served, it will be important for the transmissions of other branches of the Hierarchy of liberated masters on Earth to be offered into the collective space also, with a similar analysis of their unique gifts and limitations. It is only by doing this that the wisdom present in all the multidimensional cultures of Earth can be integrated and then synthesized so as to inform the next collective stage of our engagement with Reality and the multidimensional kosmos.

The Three Phases of the Teaching

The Trans-Himalayan teachings as a whole are understood by many to have three primary phases of expression, the last of which is emergent, and yet to have been fully given. The Tibetan Master Djwhal Khul, in his work with Alice Bailey, described this unfolding of the teachings over time in most detail,[10] though Helena Blavatsky also hinted in her greatest work, *The Secret Doctrine*, that there would be a continuity to their release over time.[11] The three phases are understood to embody a continually unfolding revelation concerning advanced mysteries of Reality, the kosmos and the Earth, with which humanity is now able to work.

The first of the three phases involved the work of Helena Blavatsky principally, and the founding of the Theosophical Society. It also included the work of such individuals as Alfred Sinnett, W. Q. Judge, Annie Besant, Charles Leadbeater, Mabel Collins, Francia La Due and Rudolf Steiner. This phase of work entailed one of the most globally significant exchanges between Eastern and Western paths of spirituality, cosmology, philosophy and scientific thought of the modern era. Indeed,

---

10. Bailey, 1960, p. 255.

11. Blavatsky, 1928, p. 22.

the large-scale prevalence of Eastern philosophy in Western culture today, as well as the increasing interface between science and spirituality, certainly stems in a large part from the pioneering work of these early Theosophists. That Blavatsky was also clear that the transmission she was bringing had been shared with her by a community of liberated masters—beings whose level of development far transcended anything that humanity was even aware of as possible—was a point that bore significant impact on world culture.[12]

Figure 2. The seal of the Theosophical Society—the primary embodiment of the first phase of the Trans-Himalayan teachings.

The second phase involved the work of Alice Bailey, a former theosophist who went on to work with the Tibetan Master Djwhal Khul, one of the beings with whom Blavatsky had come into contact during her phase of work. Bailey described that Djwhal Khul first contacted her telepathically in 1919, when she was 35, and communicated that she might be able to make a contribution to the evolution and awakening of humanity through their collaboration on a series of books. Bailey refused initially; she later described that she had no interest in what she perceived to be lower psychic activity and at the time did not know

---

12. Some readers may be familiar with allegations of forgery aimed at Blavatsky concerning the true source of the teachings she received being from such a community of awakened masters dwelling in the Himalayas as is here described. Should any such readers have any doubts about the authenticity of Blavatsky's claims, they should consult the eminent Theosophical scholar, David Reigle's article, "Why Take Blavatsky Seriously", which can be found at: www.easterntradition.org/why%20take%20blavatsky%20seriously.pdf

how trustworthy this mysterious contact was. He persisted, however, and after a trial period in which she was allowed to weigh up whether what came through really was genuine and of service to humanity, she agreed to take the dictation of his teachings. Thus began a telepathic relationship that continued for 30 years until Bailey passed over in 1949.

Alice Bailey was always clear that the form of communication she shared with Djwhal Khul was telepathic rather than channelling. Bailey noted that she did not 'give up' her mind in the process, but rather was trained to align it sufficiently with that of Djwhal Khul's consciousness and energetic transmission for the teaching to flow through cleanly and unelaborated. Together they collaborated on 24 books of esoteric philosophy and inner science, while Alice and her husband, Foster Bailey, established the Lucis Trust. The Lucis Trust still holds under its umbrella the Arcane School—one of the most respected esoteric schools of meditation, study and service in the world; Lucis Publishing, who continue to publish the Djwhal Khul/Alice Bailey teachings; and World Goodwill, a global educational initiative that seeks to promote right human relations through all domains of human activity.

Figure 3. Symbols associated with the works of Alice Bailey (left), Helena Roerich and the Agni Yoga material (center) and the Lucille Cedercrans material (right)—the three major expressions of the second phase Trans-Himalayan teachings.

The second phase also involved the work of other individuals such as Helena Roerich and her husband, Nicolas Roerich. Helena worked with another of the Trans-Himalayan community of masters, a being known as the Master Morya, on what are known as the Agni Yoga teachings. Her husband Nicolas also partook in this reception of teaching, and remains one of the most respected spiritually influenced artists of the 20th Century. Lucille Cedercrans is another woman of definite note to the second phase. She worked with the Master Rakoczi on what they called the "New Thoughtform Presentation of the Wisdom" in much the same manner that Bailey worked with Djwhal Khul and Roerich worked with Morya.

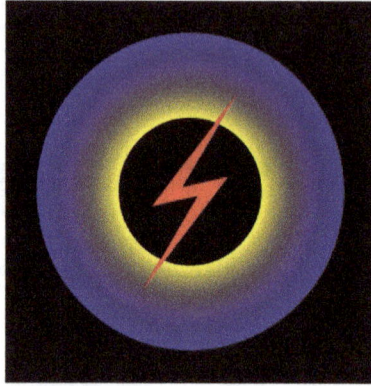

Figure 4. The Shamballa School symbol—an expression of the emerging third phase of the Trans-Himalayan teachings

In his work with Bailey, Djwhal Khul suggested that the third phase of the teachings would emerge around the year of 2025,[13] but that teachings preparatory to the third phase would emerge in the early part of the 21st century.[14] In his work with Bruce Lyon, Dwjhal Khul has suggested that the third phase of his teaching will have the Boundless Immutable Principle of the One Life, or what following Tibetan tradition we also describe in this book as Awakened Awareness, as its primary focus, as well as the monadic depth of our relative self that exists as kosmic Life-force, and the galactic story of becoming within which humanity finds its place.[15] The preparatory teachings for the third phase began to emerge in the form of a 10-year collaboration between Djwhal Khul and Bruce Lyon between the years 2000-2010; they have been, and continue to be, of deep inspiration to Shamballa School, a community founded by Bruce in 2001.

In terms of the full third phase teachings that are projected to emerge around the year 2025, it is understood that the new teaching will not be given through any individual, but through a trans-lineage

---

13. Bailey, 1960, p. 255.

14. Bailey, 1955, p. 261.

15. Lyon, 2010, p. 13, 17, 59, 171, 212. See also Lyon, 2003, p. 29-37.

group initiate; in other words, souls who have penetrated with stability into the Infinite Awake Light beyond any tradition and who recognize each other as unique expressions of that Light.

I would suggest that this group initiate will not by any means be composed of individuals who all have a pre-established connection to the Trans-Himalayan teaching. Rather, it will be composed of individuals from many traditions who have entered into spheres of radical and evolutionary awakening that have taken them beyond their traditions into the universal space of identification with Absolute Reality, the one humanity, and the sacred kosmos. This relates to the central circle in the graphic shown in the previous chapter (Figure 1). Only such a group would be able to represent a synthesis of global perspectives in the cultures and lineages they have grown to transcend and include, and thus open an emergent door to that which is to come. For such beings, the choice to stand together as one will come not from shared systems of belief or methods of work, but because it is the irrefutable will of their free and awakened hearts to do so.

Developmentally, such a group will likely be composed of beings who are both radically awake to and as Awakened Awareness—the Boundless Immutable Principle—and evolutionarily awakened to the personality, soul and monadic levels of their being. Their radical awakening will allow them to stand as the divine Freedom and Heart-Radiance of Reality that is no less present in humanity than any other kingdom. Their evolutionary awakening would simultaneously preserve the dignity of humanity's place in the great chain of being by allowing us to more deeply serve the electric transmission of planetary Purpose emanating from Shamballa. As such, my sense is that radical awakening will serve as the basis of the third phase teaching, in terms of the state of awareness that it is transmitted from, the state in which it is received, and in terms of its mysteries forming a foundational pillar of content. Additionally, from an evolutionary perspective, such a group will have as its focus the revelation of the mysteries related to the monadic level of our being and the planetary Purpose of Earth.

In terms of the purpose at the heart of this third phase of teaching that is to come, it is understood that this is analogous to the ultimate destiny of the soul depth of our being. Over its great cycle of untold incarnations, it is the role of the soul to stand between and mediate the relationship between what the Trans-Himalayan teachings call the personality and monad until, at an advanced stage of development, such dynamic fusion occurs between the two that the soul is consumed in the flames of its own sacrifice. Correspondingly, so also is it said that the Trans-Himalayan teachings *are already that which they seek to teach*—a space of communion both between multiple fields within humanity, but also between humanity and other, super-conscious kingdoms of life. As such, the intention of the teachings is to bring Humanity, the Earth, and the Fire of Reality and planetary Purpose transmitted by those super-conscious communities into such fusion, that the movement is eventually consumed in the flames of a *global* awakening, having realized it's destiny. As Djwhal Khul describes it,[16][17] this realization of its essential purpose will contribute to an even greater revelation beginning to break like a wave upon humanity's consciousness and the whole of the Earth. This is said to be a revelation concerned with the Ultimate Reality of Awakened Awareness realized and expressed in a global culture of the One and though a universal spirituality. It involves humanity's identification as a single being of kosmic origin and significance. And it entails our realization of the sacred Purpose that it is the responsibility of the Earth to transmit in the kosmos, once every particle of matter here is radiating as recognized Godhead.

This relates to the ultimate destiny of the Trans-Himalayan teachings to be offered up like a rock of incense lit on the altar of a universal spirituality, and for its body to burn away and leave only the fragrance of its essence. This perfume, once liberated, can then inform the trans-lineage space out of which a universal spirituality can be born. This speaks to the recognition that the Trans-Himalayan teachings were never actually intended to be another pathway or spiritual tradition. They have a different destiny, and are only designed to act as a conveyor belt for souls through the stages of awakening and growth temporarily, whereas other traditions will remain in place as conveyor belts of evolution. Ultimately, the Trans-Himalayan movement's purpose is to journey headlong into its own end; to act as a match, struck by its students' awakening to Reality and contact with the Life principle and then thrown into the kindling created by both its previous phases and the other traditions. The students of the Trans-Himalayan teachings are thus called to penetrate like a single arrow straight into the universal spirituality space, but to wipe their feet as they go, and to not leave any footprints. That is, we do not leave a pathway behind us.

In the third phase of the Trans-Himalayan teachings it is understood that this will initiate the disappearance of the movement from the world. No dimensionally defined form is destined to last forever, and what better way to go out than in the Fire of realized destiny? Indeed, it is this transmission of Eternal Unity and planetary Purpose that abides at the heart of the entire Trans-Himalayan transmission, and it

---

16. Lyon, 2010, p. 12.

17. Bailey, 1960, p. 257.

is more important than any of the words or concepts through which it is expressed.

In one of the books published as a result of this recent collaboration between Djwhal Khul and Bruce Lyon, *Occult Cosmology*, the Trans-Himalayan teachings are reviewed in each of their phases, and it is suggested that each phase can be understood to play a particular role in the awakening of those who work with them.[18]

The first phase, which expressed itself in the Theosophical movement, is described as *orienting* the individual to the spiritual realities, and this can be clearly seen to be the case with the work of the early and present day Theosophists. The Theosophical movement was and is primarily one in which the communication of 3rd person information plays the predominant role.

The second phase is described as providing a body of teaching that allows the individual to learn to *participate* in the evolution of consciousness and the working out of the sacred process of evolution and awakening that is occurring on Earth. This can be seen in the incredibly detailed picture of the evolving planetary and solar ecologies given in these teachings within which we find our place, and the methods whereby we can enter into 2nd person relationship with that process.

The keynote of the third phase is described as incorporating a set of teachings whose power may allow the full liberation of the individual. It is suggested that from their 1st person radical awakening Awakened Awareness, they may begin to consciously work with and powerfully transmit kosmic currents of creativity, love, wisdom and divine will; that they can do this in enlightened relationship with cultures of all kingdoms, both pre- and post-human; and that the spectrum of their service can span multiple planes.

The planetary culture and civilisation of Reality that is herein envisioned will likely not fully manifest for some time, and yet, with the need of the world and the crises that humanity currently faces so great, we feel that the call toward the manifestation of our planetary Purpose is fundamentally important to sound. Furthermore, it seems additionally true that the current world crisis also provides such a field of intensity that our collective awakening and evolution have the opportunity to flower with a speed never before known.

Even within the long-arc context of the unfoldment of our planetary Purpose, we believe that the universal spirituality to which the Trans-Himalayan teachings have contributed is something that can begin to emerge *now*. Large numbers of human beings, both within and outside of traditions, are awakening to Absolute Reality and moving into worldcentric spheres of growth. Many now stand ready and open

---

18. Lyon, 2010, p. 12.

to move into a space of trans-lineage communion where the gifts that their lineages and backgrounds are able to bring can be essentialized, synthesized, optimized and translated into new, universally accessible forms.

From one angle, Dustin and I see this book as an opportunity to engage this process specifically with the Trans-Himalayan teachings. As such, we see it as a potential bridge for Trans-Himalayan students into this universal space. And we also see it as a gift to those from other backgrounds and traditions who are already inhabiting it and who may be interested to receive what the Trans-Himalayan teachings are holding.

As an offering to this process, to humanity, the single tradition that is our planetary awakening, and the One Reality that is Infinite Awake Presence, the best synthesis of the Trans-Himalayan teachings that I am currently able to present is herein shared.

Chapter 2 - The Trans-Himalayan Teachings

Part 2: The Self

There are the two Paths by which souls may Realize the Eternal Bliss that is Truth, the Condition of the Divine Person. One Proceeds by gradual ascent of attention within the planes of experience, dissolving, by degrees, the coverings of the soul in the planes of illusion and change, until the Heart Out-Shines the inner being. The other Path Proceeds directly and immediately to the Heart, the Radiant Self, prior to all experience, all progress, all strategic austerity, all dramas of attention. One Path Proceeds by mystical ascent of the illusory inner self, or mind, via the Chain of Creation. The other Proceeds by direct intuitive submission of the entire body-mind into the Radiant Source and Transcendental Matrix of all phenomena.

(Adi Da)[19]

---

19. Adi Da Samraj, 1978, p. 107.

Chapter 3
Radical Awakening and Evolutionary Awakening
Jon

One point that is important in the most recently emerging Trans-Himalayan teaching is the difference between two fundamental vectors of spiritual unfoldment—radical awakening and evolutionary awakening. The distinction between these two forms of awakening was not present in the first two phases of the Trans-Himalayan teachings (Theosophy and then the teachings that came through Alice Bailey and Lucille Cedercrans), and only emerged through the third phase teachings from Djwhal Khul to Bruce Lyon. We feel this differentiation is crucial to a deep understanding of human potential as a whole, and how that might be honored in a universal spirituality.

The combination of these two forms of awakening is increasingly central to the newest and most synthetic forms of spirituality emerging today, from Ken Wilber's distinction between growth through state-stages and structure-stages, which he also speaks of in relation to the realization of our True Self, and the actualization of our Authentic Self; Adi Da's teaching on Transcendentalist and Emanationist paths; Andrew Cohen's focus on Being and Becoming in his Evolutionary Enlightenment teachings; and Thomas Hubl's focus on Silence and Movement. Here is how Bruce Lyon defines these two forms of awakening:

> Evolutionary awakening refers to the process of initiation wherein self-conscious individuals gradually and sequentially experiences themselves through transformation becoming identified with subtler and more inclusive levels of identity. Radical awakening

occurs when self-conscious individuals suddenly and radically experience themselves as the One Life.[20]

As is suggested in this quote, radical awakening is awakening to and as Absolute Reality, or what is described as the Boundless Immutable Principle in the Trans-Himalayan teachings, which transcends, includes and is arising as the entire kosmos. This Absolute Reality is described variously in the traditions and in this book with such terms as Awakened Awareness, the Absolute Self, the One Life, Infinite Awake Presence, Source, or Ultimate Reality, for instance. This realization has been cultivated in the various radical awakening lineages of Earth such as Dzogchen, Mahamudra, Zen, Vedanta, Sufism, and Kashmir Shaivism, as well as by many individual realizers not affiliated with any tradition through time.

Evolutionary awakening refers to the shifts of a relative self's level of polarization that take place within the planes of the multidimensional kosmos, from what we will go on to describe as the personality to soul to monad bodyminds, and then into wider and deeper spheres of incarnation along the kosmic paths. Evolutionary awakening also incorporates various other lines of unfoldment too, as we shall explore in this book. These include an individual's cultivation of relationship with cultures of beings operating in all kingdoms; access to a greater number of planes of experience and their specific forms of energy and information; and the unfoldment of the various intelligences of each plane.

As we look to the forms of wisdom that have been most especially cultivated on Earth during the last few thousand years, it has often been the case that the wisdom on radical awakening to Absolute Reality has been more primarily cultivated in the East. This does not discount the fact that the West has of course had its exemplars of radical awakening, such as Plotinus or Meister Eckhart for instance. Nor does it disregard such research as that presented by the scholar-mystics, Scwhaller de Lubicz and Peter Kingsley, among others, which points to Ancient Egypt and Greece having had their own thriving radical awakening traditions. Rather, it simply acknowledges that the meticulous level of detail with which awakening to Absolute Reality has been described and preserved in the Eastern lineages for at least the last two millennia bears testimony to its special rooting there. Conversely, though the East has its notable theorists on evolution such as the Taoist philosopher Zhuangzhi, and the Indian sage Aurobindo, it can also be seen that the understanding of evolution has been primarily unpacked in the West. This has included the contributions of such theorists as Hegel,

---

20. Lyon, 2010, p. 55.

Marx, Darwin, and more recently through such authorities as Richard Dawkins, Jürgen Habermas, and Ken Wilber.

Interestingly, this radical/evolutionary division has also qualified the first and second phases of the Trans-Himalayan teachings. The first phase, expressed primarily through the Theosophical Society, offers a cosmology that gives detailed teachings on the arising of Absolute Reality as the manifest multidimensional kosmos. The teachings of the second phase, as presented through Alice Bailey, Lucille Cedercrans and Helena Roerich, were on the other hand far more explorative of evolutionary awakening within a deeply sophisticated kosmocentric context.

I would suggest that this differentiation of emphasis is no coincidence, and that it is actually expressive of a fundamental point of integration between radical and evolutionary perspectives that the Trans-Himalayan teachings are ultimately intended to pioneer. I also propose that the third phase of the teachings, when it emerges fully around 2025, is planned to involve a true synthesis of these radical and evolutionary perspectives in a manner that:

a) Opens a space for them to spread increasingly through humanity as a whole, both within and outside of the boundaries of the traditions.

b) Empowers humanity's understanding of itself and the Earth within a kosmocentric context.

c) May allow humanity to significantly augment its collective apprehension of advanced stages of radically awake evolutionary unfoldment, such as those which unfold in post-mastery stages of development.

As such, this book seeks to present an understanding of the spiritual path and our deepest potential that includes both the Ultimate Reality of the One Life revealed in radical awakening and the never ending path of evolution that extends from Earth deep into the multidimensional kosmos. Indeed, from one perspective, this book could be considered an in-depth exploration of these two forms of awakening; how they differ from each other, where they overlap, and how they interact on the path. It is our hope that this will be of service to those brothers and sisters around the world for whom these matters are increasingly relevant and revealing themselves in their own experience.

Chapter 4
A Dual Center of Gravity
Dustin

As introduced in the previous chapter, we recognize that while the Absolute Self has no developmental trajectory or dimension, the relative self travels along two vectors of growth: a vector of radical awakening and a vector of evolutionary awakening. Most traditions tend to focus on only one of these vectors while ignoring the other. In our work we honor both, and suggest that such a position will be fundamentally important for a holistic, integral, universal spirituality.

When we take both radical and evolutionary awakening into consideration, the self can be understood to have a radical center of gravity and an evolutionary center of gravity. In the language of Integral Theory, we call this a state-stage center of gravity (corresponding to the path of radical awakening) and a structure-stage center of gravity (corresponding to the path of evolutionary awakening). When these two stations of the self are taken together, Wilber calls this a "dual center of gravity".[21]

We sometimes refer to the development of self along the radical and evolutionary awakening vectors using spatial orientation. Horizontally, every self has what Daniel P. Brown calls a vantage point,[22] or what Ken Wilber calls a state-stage identity unfolding along the radical awakening vector. Vertically, every self has an *altitude*, or a level of developmental unfolding along the evolutionary awakening vector. Both spectrums of development follow the pattern of increasing *envelopment*, which means that each new stage transcends and includes the former.

---

21. Unpublished work. Personal Conversation.

22. "Vantage Point" is a term first introduced by Daniel P. Brown.

At every stage of development, both radically and evolutionarily, the self serves two roles. The first is as a *constant function*, in terms of its role in serving as the locus of integration for all the disparate aspects and streams of unfolding. The second is as a *developmental stream*, or line of unfolding, in itself.

In relation to the first of these, the self as a constant function of integration, it is this function that maintains a healthy locus of integration between all the various lines of unfoldment. As we will be considering these lines in this book, these are the lines of identification, experienced relationship, plane access, and a collection of intelligence related lines.

In relation to the self as a line of development in itself, this naturally unfolds along the two vectors of radical and evolutionary awakening, to produce the self's dual center of gravity. The radical awakening vector relates to the path along which the relative self undergoes a total deconstruction so as to realize itself as unbounded, Nondual Awakened Awareness. And the evolutionary awakening vector entails a continual abstraction into deeper and ever wider spheres of incarnation in the kosmos.

A core finding for developmental researchers relevant to both of these orientations of development, though in more conventional forms, has been the needed distinction between two aspects of the self-sense. The first is that interior and subjective awareness that many people sense as if located just behind the eyes, which is the witness of everything occurring in our world. This has been called the *proximal self* (proximal because it is 'nearest' to our experience), and it relates to our sense of 'I'. The second is that aspect of the self that is witnessed by the proximal self. This has been called the *distal self* (distal meaning 'more distant'), and can be understood according to our sense of 'me'. The importance of this distinction rests in the finding that as the process of development unfolds, both radically and evolutionarily, the proximal (or most intimately 'near' witnessing self) becomes the distal, or objectively-viewed, self of the next stage. In the words of Robert Kegan, "the subject of one stage becomes the object of the subject of the next stage."[23] This means that in both the horizontal vector of radical awakening and the vertical vector of evolutionary awakening, the sense of self at one stage transcends, includes, and penetrates the sense of self at the former stage.

One way that the Trans-Himalayan teachings illustrate this diagrammatically is to consider the relative self as a point of awareness (the proximal sense of self) found at the center of three lines each

---

23. Wilber, 2007, p 127. See also, Robert Keegan, *In Over our Heads*.

intersecting at 90 degrees and representing the three dimensions of height, width and breadth. This can be seen below:

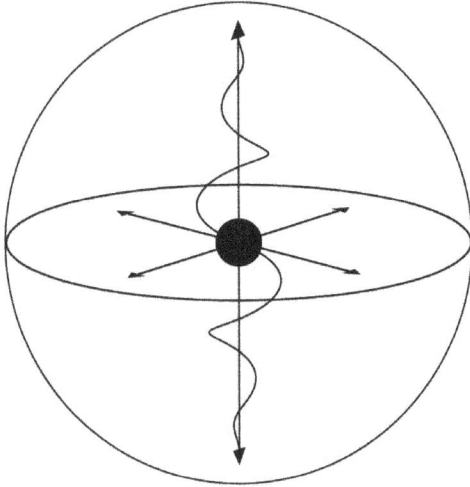

Figure 5: A diagram showing the various dimensional orientations along which awareness can move.[24]

As is shown in this figure, in terms of evolutionary awakening the point of awareness that is the relative self can spiral up and down the vertical line. This movement represents a line of development that we call identification, and which involves an evolutionary expression of the proximal sense of self at each level becoming the distal self of the next (from personality to soul to monad and beyond, as we shall go on to see). Looking to the horizontal arrows pointing outwards from the center, it can extend its awareness so as to cultivate relationship and access to different planes of experience. Or, along the radical awakening vector and as we shall see in the next section, it can turn its gaze back upon itself so as to recognize its true nature. By doing so, it comes eventually to the realization that its entanglement with thought, personality, time and even a presumed locus in space as 'individual consciousness' was only apparent, and that its True Nature is unbounded Awakened Awareness—the One Life within and as which the whole kosmos is arising and unfolding.

---

24. Reproduced from Lyon, 2010, p. 58.

From the perspective of Integral Theory, all human beings[25] are born at the shallowest vantage point (with a vantage point identified only with the Gross state of the Absolute) in the radical awakening vector. They are also born at the earliest stage of evolutionary growth (with identification centered entirely in the physical body, and the other evolutionary sub-lines of relationship, plane access and intelligence wholly centered on the physical plane) in the evolutionary vector. The graphic below shows a self with a dual center-of-gravity at the Gross state along the radical awakening vector, and physical along the evolutionary vector. Notice the circle in the lower left hand corner of the graphic.[26]

---

25. It is the opinion of the authors that this point needs deeper clarification. One who dies while maintaining recognition of Awakened Awareness carries that awareness through the intermediate state after death. If the being then decides to take incarnation in the physical plane again, recognition of Awakened Awareness can remain through the birth process depending on the being's level of stabilization. This is the phenomena described in the context of some Tibetan tulkus. However, given the fact that the relative self-sense (in the evolutionary vector) is not fully formed and differentiated until around the age 2, we feel it likely that in many cases recognition of Awakened Awareness is temporarily dormant until the differentiated self-sense emerges.

26. It is important to note that this particular graphic shows the various perspectives and directions of growth that can unfold within a single space-time continuum. Simultaneously, we can hold open the possibility that Awakened Awareness can also experience itself at varying degrees of multitemporal maturity. In these cases the single point would be existing in several spheres at the same time. This is something we explore more deeply in Chapter 9.

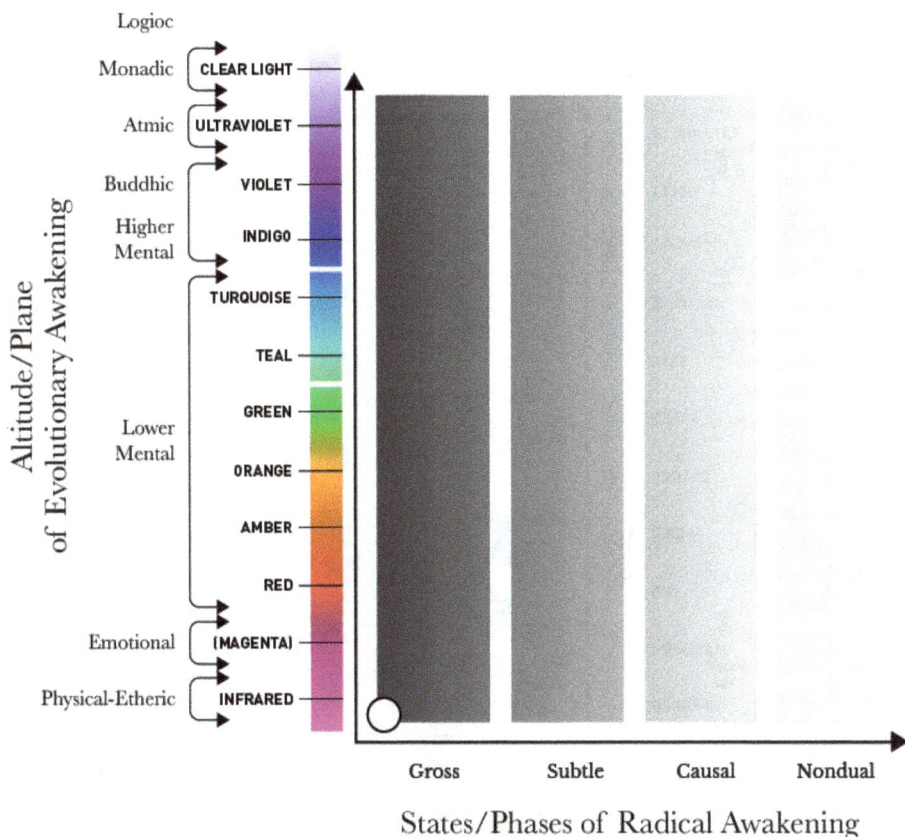

Figure 6. Chart showing the starting point for growth across
both the radical and evolutionary awakening vectors.

Over the course of development, a relative self has the opportunity
to awaken horizontally along the radical awakening vector and to
grow vertically along the evolutionary vector. The speed and pattern
by which this takes place is determined by subjective, biological-
behavioral, cultural and social influences (these are the Four Quadrants
of the AQAL matrix, an important tool in Integral Theory that we
will go on to explore more fully shortly). The graphic below shows the
life path of an individual who has awakened along both spectrums to
a dual center of gravity comprising a Subtle state/vantage point and a
mental (orange) altitude of identification:

71

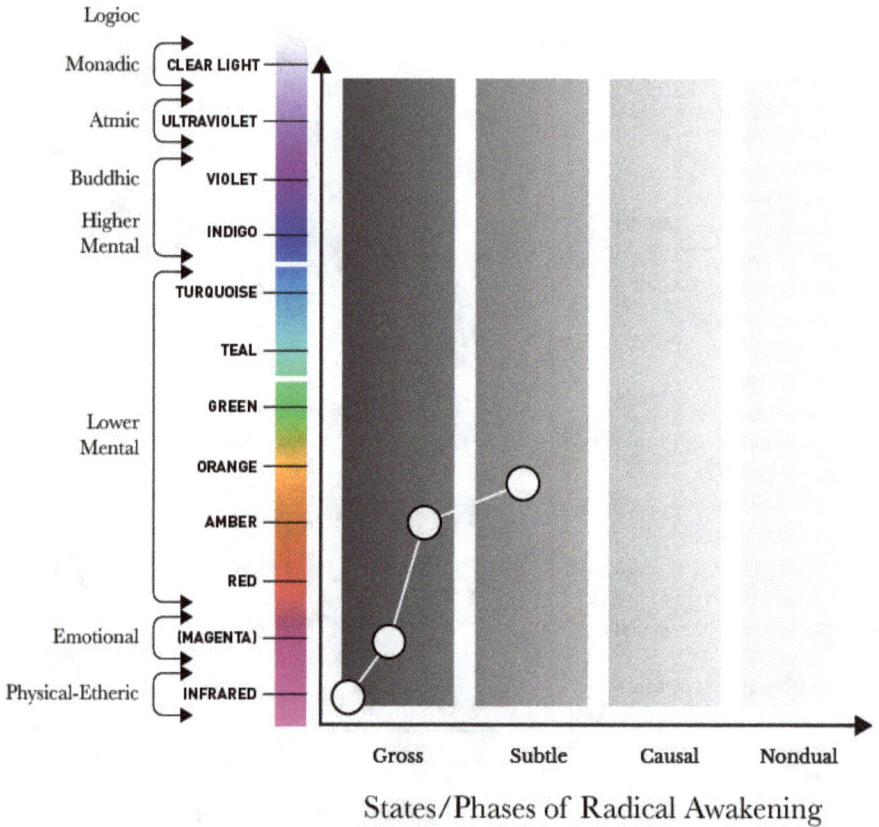

Figure 7. Chart showing an example of a possible trajectory of growth for a human being across both the radical and evolutionary awakening vectors.

We explore both of these processes of awakening and growth in full detail in Section 3 (on Radical Awakening) and Section 4 (on Evolutionary Awakening). For now, it suffices to say that both the Trans-Himalayan teachings and Integral Theory consider it vital to include both a vertical and horizontal dimension of growth in order to understand and empower human beings' path into their fullest potential.

Chapter 5
The Four Quadrants
Dustin

The Four Quadrants of the AQAL (All-Quadrant, All-Level) matrix form one of the centrepieces of Ken Wilber's Integral Theory, which we shall be drawing upon often throughout this book and using at various scales of the kosmos. To explain the Four Quadrants we can start by noting that every self, including you, emerges as a confluence of four perspectives. These perspectives, as Wilber points out, include an interior and an exterior as experienced from both the individual and collective dimensions of life. This means that as the self awakens and develops through the vectors of radical awakening and evolutionary awakening, all Four Quadrants *tetra-mesh* and *tetra-arise* as an inherent part of reality. That is to say, every experience can be viewed from one of these four perspectives no matter where a being might be on the developmental spectrum.

According to Integral Theory, the Four Quadrants arise holonically. Holons are theoretical constructs used by such theorists as Arthur Koestler and Ken Wilber that are proposed to constitute the whole of reality. Every holon is understood as both *a whole in itself* as well as a *constituent part* of a greater, more encompassing whole. One example would be a letter in a word, a word in a sentence, a sentence in a paragraph, a paragraph on a page, and a page in a book. Another example might be an atom within a molecule, a molecule within a cell, or a cell within an organism, etc. In all instances, holons are both wholes in themselves and also parts of greater wholes. In accordance with this definition, nested spheres of such relationships on greater and greater scales are termed a *holarchy*.[27] Every holon which arises in the

---

27. See Wilber, *Sex, Ecology, Spirituality*, p 22 for a discussion on holarchy.

kosmos has Four Quadrants, all the way up and all the way down.[28] Now, let us explore each perspective one at a time.

The Upper Left Quadrant (UL)

When examining your own experience, the Upper Left quadrant, which we abbreviate simply as the 'UL', relates to the subjective phenomena that arise in the interior of your own awareness. For instance, the interior of your experience includes your thoughts, emotions, intentions, your sense of identity, or anything else that is uniquely yours and yet not observable in the exterior world. Right now, as you read, notice what is happening in your own awareness. What thoughts and emotions are arising? What are the phenomenological characteristics that you can see in your own experience that are not observable to others? This is your Upper Left quadrant.

The Upper Left quadrant, as examined in one's own direct experience, is the domain into which most spiritual, religious, and contemplative traditions have offered their wisdom. Radical awakening, as it develops through various vantage points or what Integral Theory describes as state-stages (moving from Gross to Subtle to Causal to Nondual), is tracked from inside the experience itself. This is the same along the evolutionary awakening vector,[29] and particularly its identification sub-line (interior awareness of shifts in relative identity from the personality bodymind to soul bodymind to monadic bodymind and beyond).[30] Usually within a

---

28. For a detailed discussion of how we can understand the Four Quadrants to exist for holons as small as atoms, for instance, see Wilber's *Sex, Ecology, Spirituality*.

29. It is important to note that, because of its specific focus on structure-stages in the mental plane, classical Integral Theory does not hold that the interior of the evolutionary vector can be tracked. The Trans-Himalayan teachings suggest, however, that evolutionary development (from personality to soul to monadic bodyminds) *can* be known directly through examining the Four Quadrants of your experience as it arises through different planes at any time.

30. Throughout this book, we suggest that even though shifts along both the radical awakening vector and the identification sub-line of the evolutionary vector are tracked from the inside the experience (UL), there is an important difference. Shifts through state-stages or vantage points of Absolute Reality progress through recognition of the whole of reality in its various phases of creation without giving any particular preference to certain characteristics. Shifts through the identification sub-line of the evolutionary vector involve movements through environments where there *is* emphasis given to particular characteristics. For instance, at the soul level the

lineage, individual spiritual practitioners compare their own first hand experiences with each other and the tradition at large. Over time, these self-reports are tracked, categorized, and synthesized. As a result, we inherit elaborate models of spiritual development that outline the stages of spiritual growth.

The Upper Right Quadrant (UR)

In addition to your own experiences on the interior of your awareness in the Upper Left quadrant, all sentient beings also have exterior bodies as pointed out by the Upper Right quadrant, often abbreviated simply as the 'UR'. One way to give an example of this is to point out that each level of identification (personality, soul, monad) has a corresponding body/form/sheath, existing in increasingly subtle frequencies of energy-matter, which it uses to interact and engage with others in the world. Identification as the personality corresponds to the use of physical, etheric, astral/emotional and lower mental bodies. Identification as the soul corresponds to the use of the soul lotus body and the higher mental, buddhic and lower atmic sheaths. Identification as a monad corresponds to expression through the higher atmic, monadic and logoic bodies. In the Upper Right Quadrant we find the space in which each of these bodies, as well as their constituent elements (such as atoms, molecules, cells, organs, a reptilian brain stem, limbic system and neocortex, for the physical body, and the subtle correspondences to these in the subtle bodies such as the etheric, astral, mental, and so on) are found.

The Lower Left Quadrant (LL)

Individuals are always part of a collective. As such, every sentient being also has collective perspectives that arise as part of experience. Just as every individual has an interior and exterior dimension, so too every collective has an interior and exterior.

---

consciousness, energy, culture and plane has particular qualities, each of them saturated with love and wisdom. On the other hand, the subtle phase of creation, there is no perceptual bias toward any of its particular aspects. Rather, there is merely a sense of the entire Subtle layer of Reality as it emerges from the Causal and comes into Gross form.

The Lower Left Quadrant, often simply abbreviated as 'LL', refers to that part of your experience that includes all inter-subjective realities. As such, it includes all your shared values, morals, perspectives, and places of emotional, intellectual, and spiritual overlap that allow you to create relationships. This is also the perspective of your being where all norms and culture arise. We often refer to the LL quadrant in terms of 'planetary culture', or in terms of 'relationship'.

Just as identification and form/body track along a spectrum from personality to soul to monad, all inter-subjective realities of the Lower Left Quadrant can also exist on the personality, soul and monadic planes. At the physical, emotional and mental altitudes this aspect of being may be expressed by the shared values, or culture, created by a community of personalities.[31] In a global sense, this is Humanity. At the soul level (the higher mind, buddhi and lower atmic altitudes), the Trans-Himalayan teaching understands the Lower Left to be expressed in the form of the planetary meta-sangha that is Hierarchy. This is the shared planetary culture of the global community of awakened and liberated souls—beings from all traditions and none who have journeyed into post-liberation stages of becoming. And on the monadic level (the higher atmic, monadic, and logoic planes), a planetary culture constituted of monadically awakened and identified buddhas is understood to exist. This planetary culture is composed of beings as far advanced in their evolutionary unfoldment as the awakened souls of Hierarchy are of humanity. This is Shamballa, within which buddhas of vast realization and evolutionary advancement hold identification with and transmit the kosmic Will, Purpose and Life of the Earth into the various kingdoms of the planet.

The Lower Right Quadrant (LR)

Finally, all beings have an exterior to the collective aspect of their being. In this way, all occurrences include a networked system of social factors and energies. As such, the Lower Right Quadrant, often abbreviated simply as 'LR', comprises the physical to increasingly subtle planes of energy-matter upon and within which our interior altitude of identification (UL), the particular

---

31. As is explored in Chapter 11, collective cultures can exist on any plane and at any scale. This perspective holds open the reality that cultures are not only personality-based or human-centric.

body through which it expresses (UR), and the culture of which it forms a part (LL) reside. For instance, the physical plane exists as the collectively shared physical environment within and upon which all physical activity expresses. The astral plane serves as the shared matrix of subtle energy-matter within which all emotional activities occur. The mental plane serves as the shared field of even subtler energy-matter within which all activities of thought and mental intention occur, and so on.

Furthermore, within each of these planes, there may be collective sub-systems through which the life of that plane expresses. As an example, on the physical plane we find economic, social and political systems, as well as both physical and social sets of laws that govern action in a particular location, through which humanity interacts with itself. The physical plane also contains within it the fields of expression for the various kingdoms, or evolutionary life-spheres (mineral, plant, animal, human), that are expressing upon it. The Trans-Himalayan teachings, like many other esoteric cosmologies, understand this to be a phenomenon not specific to the physical plane. On every plane they understand there to exist various different forms of life and consciousness.

The higher mental, buddhic and lower atmic planes include, among other things, the entire matrix of collective energies that arise between souls—this is the collective energetic field of Hierarchy. Just as the physical plane has its own sets of natural laws, so too the buddhic and atmic planes have their own natural laws that govern action in a particular location. All planes do. Subtler than this, the higher atmic, monadic, and logoic planes are home to the inter-objective matrix of monadic energies, monadic environments, monadic laws, and the collective energetic field of Shamballa.

The graphic below offers a representation of Wilber's Four Quadrants that demonstrates the model we are describing. Notice the graphic shows the interior and exterior perspectives of the individual and the collective. Although the graphic below only shows one cross-section of reality (it does not show the different levels, or altitudes of being, as they arise in each quadrant), it should be clear from the descriptions above that all Four Quadrants tetra-arise in all planes of reality (physical-etheric, emotional, mental, buddhic, atmic, monadic, logoic, etc.).

	Interior	**Exterior**
Individual	UPPER LEFT (UL) Identification	UPPER RIGHT (UR) Body/Form
Collective	Relationships/Culture LOWER LEFT (LL)	Collective Energy Fields LOWER RIGHT (LR)

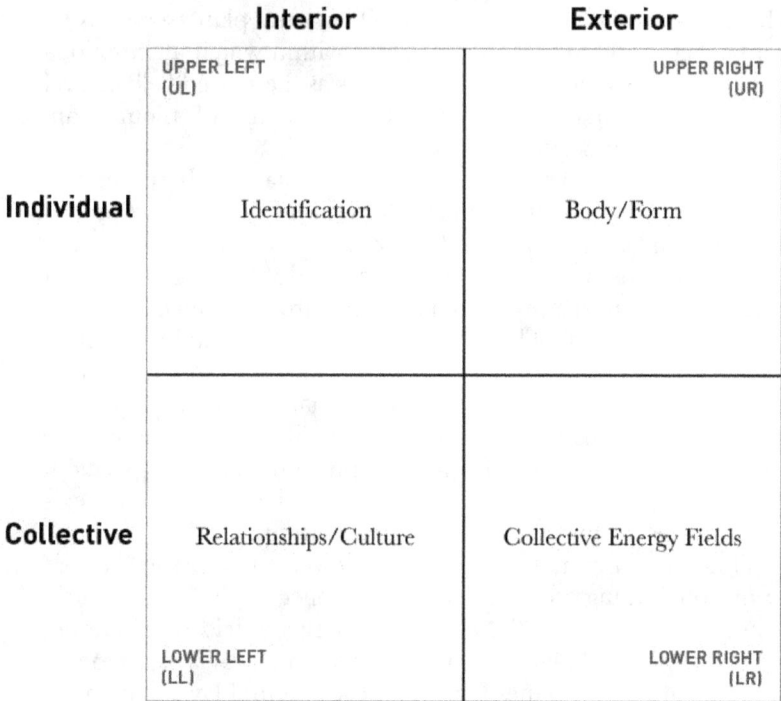

Figure 8. The Four Quadrants: Identification (UL), Body/Form (UR), Relationships/Culture (LL), Collective Energy Fields (LR)

As we explore together, we use the Four Quadrants throughout this text to point out identification (UL), body/form (UR), the interior sphere of relationships or culture (LL) and the exterior energy fields of those planetary cultures and the environments or planes on which they reside (LR). Although we give more emphasis to how these Four Quadrants show up along the evolutionary vector, we also point out several ways that they all simultaneously arise as the self navigates the radical awakening vector as well.

Chapter 5 - The Four Quadrants

Part 3: Radical Awakening

In the stillness of the night, the Goddess whispers. In the brightness of the day, dear God roars. Life pulses, mind imagines, emotions wave, thoughts wander. What are all these but the endless movements of One Taste, forever at play with its own gestures, whispering quietly to all who would listen: is this not yourself? When the thunder roars, do you not hear your Self? When the lightning cracks, do you not see your Self? When clouds float quietly across the sky, is this not your own limitless Being, waving back at you?

(Ken Wilber)[32]

---

32. Wilber, 2000a, p. 279-280.

Chapter 6
The Nature of Radical Awakening
Dustin

The term radical awakening relates to the deconstruction of relative reality to its Absolute Base. This is the process by which the relative self is seen through as a mere construction and Awakened Awareness realizes itself to be none other than the Absolute Nondual Self. This is the path taught in such lineages as Dzogchen, Mahamudra, Zen, Vedanta, Kashmir Shaivism, Sufism, and Tibetan Bon. Realization of one's own native awareness as the Absolute Source and Suchness of all relative reality instantaneously gives way to a full experience of unbounded wholeness and unconditioned liberty. One is left with total satisfaction in Reality just the way that it is: a natural great perfection.

Recognition of Awakened Awareness comes in an instant. This type of "sudden" realization is said to be possible precisely because the condition of the Absolute is always-already the case. There is nothing, other than the lack of recognition itself, which stands in the way of its full flourishing. In this sense, coming to know the all-transcending, all-inclusive and all-emanating Infinite Awake Presence of the radical awakening vector is possible any time, and has no prerequisites.

Most of us, however, will not simply stumble upon awakening as a sudden realization without particular conditions being set first. Conditions that help to set up an even greater possibility for realization are referred to as aspects of what some traditions call the "gradual" path. For our purposes here, we examine radical awakening along the gradual path so as to offer the reader some sense of how the process of realization can unfold in stages.

One of the first points to make clear is that radical awakening is not simply a fleeting peak experience that comes and goes. Rather, when it is authentic, radical awakening involves the tacit and innate knowing that this is awakening to the true nature of Ultimate Reality. In this way, the realization and recognition of it can be ongoing and sustained in all times and situations. Radical awakening to the Absolute, and its continual recognition, relate to the aspect of spiritual development that most people refer to when they use the term Enlightenment.

In order to provide a well-rounded introduction to radical awakening we contemplate it from two perspectives. The first takes a consideration of the states/phases of awakening. The second looks at the source of awareness as it appears to shift vantage points. It is important to note that in doing this, we explore both states/phases and vantage points from the perspective of the gradual path rather than the view of a sudden path.[33]

We examine radical awakening from two different orientations. The first is a universal or ontogenetic orientation. From this perspective we consider the phases of universal creation as generally understood by the world's wisdom traditions. The process of creation moving from Absolute Reality into relative form is generally referred to as *involution* (Wilber, Aurobindo, etc.). The second orientation is from an individual/microgenetic perspective. This means that we introduce the basic notion of realization as it relates to awareness moment-to-moment. From this perspective, realization is a process by which one travels back along the spectrum of involutionary creation, tracing its emanatory line back to its Source through what are described as *vantage points*. The gestalt of these two perspectives taken together (ontogenetic states/phases of universal creation and microgenetic vantage points of awareness moment-to-moment) provides a sufficient introduction to this vector, and will set the foundation for us to examine a more detailed cross-cultural model of radical awakening in the following chapter.

Phases of Involution

The term involution refers to the process by which the Boundless Immutable Principle of Awakened Awareness emanates itself into relative reality. Although the specific phases of universal manifestation and unveiling and the various degrees of granularity described in the traditions tend to vary, Integral Theory points out that the process of involution, in the broadest sense, usually follows a common path. The involutionary path begins with the Absolute as Nondual Awakened Awareness, and then steps down into *Causal*, *Subtle* and *Gross* states of Absolute Reality respectively. Here we wish to reiterate that throughout

---

33. There are a number of radical awakening teachings that focus on sudden radical awakening without any focus on stages. One example here is Dzogchen.

this book we use the term "Subtle" in two different contexts. When referring to the Subtle state of the Absolute as it is relevant to radical awakening, we use a capital "S". When using the word in a more general manner and in relation to the relative evolutionary kosmos, it is uncapitalized.

The process of involution can be thought of using at least two different scales. First, it can be conceived of from a macro-universal scale. From this particular orientation one can conceive of the process by which the entire kosmos is created (kosmo-genesis/ontogenesis). The Nondual Suchness that is Awakened Awareness (the very Base and Source of Reality) first coagulates into the Causal state. This Causal formlessness is the matrix from which space and time are born.[34] It could even be said that the Causal state is the context from which space and time emerge. Causal formlessness, which is infinite unmanifest Being resting as pure potentiality and intention, then steps down into the Subtle state. This Subtle state is composed of a vast expanse of amplitude-modulated energy states that we might describe as sound, light and vibration. Subtle Reality then steps down into the Gross state that is composed of all universal Gross forms and objects, at which point in the temporal arc of the universal creation process, the Big Bang explodes into existence. Models of involution, like the one just provided, are commonly found within esoteric traditions of both the East and West.

Rather than using the macro-universal scale, as described above, the second way to conceive of involution is on a moment-to-moment scale. From this perspective one can perceive involution in *every single instant* as spontaneously present. In each moment, Ultimate Reality remains unmoved as the Clear Light of Awakened Awareness. In the very same instant, Awakened Awareness emanates as Causal formlessness. Simultaneously, Awakened Awareness is spontaneously present as the light, energy and vibration of Subtle Reality. While at the same time, Awakened Awareness is spontaneously present as all Gross form. From this perspective, there is no sequential involutionary movement into form but rather the spontaneously present existence of all three phases of Absolute Reality all at once.

When first becoming familiar with the phases of involution, it can be helpful to think of the process like water (H_2O) moving through state/ phase shifts from gas to liquid to solid. In this sense, Absolute Reality (H_2O) can be seen as manifesting as Causal Reality (gas) to Subtle Reality

---

34. As we shall see in later chapters, access to Causal layers of Reality can and often are cultivated in lineages of spiritual practice. The fact that the Causal layer of Reality is beyond both time and space allows some of these individuals to access information concerning the past and future in a non-linear and non-local fashion.

(liquid) to Gross Reality (solid). As described above, this process can be seen as sequential or all-at-once. One example of how Awakened Awareness remains as such, regardless of the state or phase that it is emanating as, can be seen in the example of liquid crystallography. Research has shown that in each perceived 'transition' phase between solid, liquid and gas, a dynamic crystal structure remains integral throughout, thus demonstrating the persistence of Suchness despite phase variance.[35] All phases of relative creation, whether gas, liquid or solid, are equally water (H_2O). In a similar way, all phases of creation (Causal, Subtle, and Gross) are equally Awakened Awareness. From the perspective of state/phases, radical awakening is the process by which awareness recognizes itself as water rather than identifying as a specific state or phase of creation. When recognized as pure Suchness, clearly and without obscuration, Awakened Awareness is known directly as the very heart and source of everything in relative reality. The realization of the Absolute Nature of Existence is what is meant by the term radical awakening.

Gradual Awakening through Vantage Points

Given the understanding of the phases of involution described above, we can now turn our attention to the process of awakening itself. Gradual paths to realization, which Wilber describes as paths that move through state-stages, and Brown calls vantage points, train awareness to hold open Awakened Awareness without allowing the self to undergo a self-contraction or process of misidentification with any particular phase. In general, the gradual process of radical awakening trains an individual to clearly distinguish awareness from all Gross state phenomena. In most cases this includes a process of distinguishing awareness from the physical body and from all coarse-level thought. As a second step this usually involves the practitioner distinguishing awareness from all Subtle state phenomena. This usually includes Subtle impulses, visions, symbols, etc. These first two shifts are sometimes associated with a more conscious and mindful life during the waking state. As a third step this involves a practitioner distinguishing awareness from all Causal state phenomena. Causal state phenomena include all universal intentions, universal karmic tendencies etc. Wilber associates awakening to this level of realization with the onset of lucid dreaming. Finally, even this Causal layer is examined and awareness is distinguished from even the subtlest forms of information processing

---

35. Thanks to David Martin for this point.

that make up individual consciousness. What is left over is affirmed. Awareness is none other than pure unbounded wholeness, arising as the entire kosmos. Wilber associates this level of awakening with Nondual realization.

As one can clearly observe, the process of gradual realization along the radical awakening vector retraces the phases of involution in reverse until Awareness finally recognizes itself as purely Nondual, the very Source and Suchness of all of manifest reality.

Both of the elements described here—states/phases of involution and vantage points of realization—are two sides of the same coin. Involution is the process by which Nondual Spirit becomes "involved" with the world. Vantage points of realization are the process by which awareness successfully deconstructs all the layers of false-identification and self-contraction that prevent it from recognizing its True Nature. Considered together, these two elements point to the nature of radical awakening as we use it here in this book.

The following chapter offers an expanded cross-cultural model of radical awakening, providing a more detailed account of this particular vector of the third phase Trans-Himalayan teaching.

Radical Awakening and Shadow

Before moving on to the next chapter it is vital to note that in any discussion about development, potential pathologies can arise. At any stage of growth (whether horizontally through the radical awakening vector as discussed here or vertically through the evolutionary awakening vector as discussed in Chapter 8), various experiences of trauma, pain, as well other naturally arising pathologies, can produce points of contraction, addictions and allergies. At each transition point, there can be pieces of the self that split off or are left behind. Extending the definition originally employed by Jung, Integral Theory uses the term *shadow* to describe these split-off aspects of self.

If spiritual practitioners move through deepening vantage points in the path of radical awakening, but fail to allow each new stage to fully penetrate the previous stage, he or she will likely suffer from a series of pathologies. In some cases, where radical awakening has moved beyond Gross state awareness and into the Subtle or Causal states, but has not fully penetrated back through the layers of manifestation and into the Gross state, the person may seem like he or she is always in a bliss state without the capacity to function in everyday reality. In more extreme scenarios, a practitioner may have the sense of total disembodiment.

Here's how Wilber frames it:

> ...once you get a strong glimpse of One Taste [Radical Awakening], you can lose all motivation to fix those holes in your psychological basement. You might have a deep and painful neurosis, but you no longer care, because you are no longer identified with the bodymind. There is a certain truth to that. But that attitude, nonetheless, is a profound violation of the bodhisattva vow, the vow to communicate One Taste to sentient beings in a way that can liberate all. You might be happy not to work on your neurotic garbage, but everybody around you can see that you are a neurotic jerk, and therefore when you announce you are really in One Taste, all they will remember is to avoid that state at all costs. You might be happy in your One Taste, but you are failing miserably to communicate it in any form that can be heard, precisely because you have not worked on all the lesser vehicles through which you must communicate your understanding. Of course, it is one thing if you are being offensive because you are engaged in angry wisdom or dharma combat, quite another if you are simply being a neurotic creep. One Taste does not communicate with anything, because it is everything. Rather, it is your soul and mind and body, your words and actions and deeds, that will communicate your Estate, and if those are messed up, lots of luck.[36]

So at the very least, when dedicated to move and act in service to the whole, development through the radical awakening vector ought to unfold in a process that includes healthy transcendence, healthy inclusion, and healthy penetration. It is the sacred obligation of the spiritual practitioner to shine light on any places that shadow might lurk so that they might share the merit of their practice most powerfully and positively with all beings.

---

36. Wilber, 2000a, p. 130.

Chapter 7
A Cross-Cultural Model of Radical Awakening
Dustin

According to the most profound and sophisticated teachings on the Absolute nature of Reality, as articulated in Dzogchen, Mahamudra, Zen, Vedanta, Sufism, and Tantric Hindu traditions of the East, there is only one true vantage point of Awareness. That vantage point is without reference point or center. In essence, it is an infinitely generative, omnipresent singularity. This is the recognition that, at its most fundamental level, Awareness is inseparable from the whole sphere of continually unfolding reality.[37]

For our analysis here, we turn to the comparative work of Daniel P. Brown. Brown is unique in that his work provides an unprecedented synthesis of the Western psychological system and several Eastern maps of meditative realization. We trace Brown's work here to show how the vector of radical awakening finds a great deal of cross-cultural substantiation. It is our hope that this effort in this chapter will further validate that this particular vector of awakening ought to be taken into consideration in any deep contemplation of spiritual development.

In a ten-year study, comparing various schools of Buddhism and Hinduism, Brown's work points to the fact that there is a single underlying meditative path to radical awakening—a path that ultimately leads to

---

37. This is not to say that all the traditions agree in their entirety about the specific qualities regarding the nature of this Base but rather to say that there are certain features underlying the path that relate to how one might deconstruct the various notions of relative reality that are part of our ordinary consciousness. On this point, the traditions agree. See the cross-cultural work of Daniel P. Brown in *Transformations of Consciousness*.

the deconstruction of relative reality and the revealing of the source of Awareness as Absolute Reality—every thought, feeling, person, being, moment, and inch of the Earth and kosmos. Although Brown acknowledges the final experience of awakening will manifest in various ways according to the biasing perspectives of the tradition in which it is realized, his work is the first to study the various meditative texts from the inside (i.e. in their original language) and to confirm an underlying deep structure of radical awakening that can be applied across cultures and traditions. This cross-cultural validation of an underlying path of deep structures provides a solid foundation upon which a universal spiritual path to awakening is revealed. As such, we see it as a crucial contribution to a universal spirituality.

In his own words, Brown explains his research methodology and its results:

> ...the Yogasutras, Visuddhimagga, and Mahamudra were compared synoptically, stage by stage, to test whether there was an underlying sequence. Each tradition divided the stages of practice differently. Textual outlines of the stages proved an unreliable way to test for common structure. However, a careful analysis of the technical language used in each text proved more useful. Using this approach it was possible to discover a clear underlying structure to meditation stages, a structure highly consistent across the traditions. The sequence of stages is assumed to be universal, despite the vastly different ways they are conceptualized and described across traditions. This sequence is believed to represent natural human development available to anyone who practices.[38]

Elsewhere, Brown explains the deep structure of the single path to radical awakening in even greater detail:

> The logical order of the one path... is incredible... Each stage expresses a particular form of de-construction. For example, ordinary attitudes, behaviors, affects and self-images are first deconstructed. Then, thinking is deconstructed. Next, perception is dismantled. While observing the subtle processes behind perception, the point of observation is altered. Next, the temporal nature of information-processing is analysed, and then, dismantled. As a result, the yogi experiences some form of interrelatedness of mind and cosmos. Finally, the

---

38. Wilber, Englar, & Brown, 1986, p. 223.

activity and observational points that interfere with this interrelatedness are dismantled and enlightenment comes forth. Each of these stages depicts an episode of structural variance along the path of liberation. The stages are invariant.[39]

Our perspective on radical awakening draws directly on Brown's most recent work, and a basic overview of the major stages of release, opening and realization in radical awakening, is a useful place for us to begin.

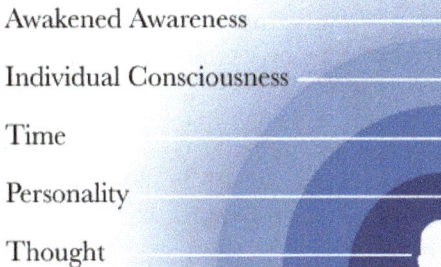

Awakened Awareness

Individual Consciousness

Time

Personality

Thought

Figure 9. Graphic showing the vantage points through which radical awakening to and as Awakened Awareness may occur along a gradual path.

---

39. See Brown's University of Chicago dissertation Mahamudra Meditation Stages and Contemporary Cognitive Psychology: A Study in Comparative Psychological Hermeneutics, Chicago, Il 1981. p 689.

For most individuals, their awareness is experienced as caught up in a multitude of differentiated relative perspectives.[40] In other words, although awareness in its most naked form is already the Absolute Base of all Reality, most individuals do not experience it as such because multiple layers of confusion and obscuration conceal its recognition. It is from the perspective of this more relative, obscured view that a description of radical awakening as a process of development is most helpful.

Most commonly, human beings assume that awareness is the same as thought (as shown in the central figure in the graphic above). As thinking flashes forth and flows in one's mind, most individuals experience no differentiation between the awareness observing the thought and the thought-stream itself. This basic assumption that awareness and thought are the same immediately conceals awareness from being recognized according to its Ultimate Nature as the single unified Whole. Confusing awareness and thought sets up a boundary within one's sense of self, and between one's sense of self and the apparent external world.

A second common obscuration is that most people confuse awareness with their personal sense of self (note the second concentric circle in the graphic above). When one's sense of self is mistaken for the personality, there is an immediate separation between self and other. Although a clear boundary of self and other is a vital stage in healthy ego development (this occurs as growth proceeds through structure-stages), it serves as a barrier to recognizing the true nature of awareness when viewed from the perspective of vantage point development. Ultimately, it is not the personal sense of self that is the problem but an *exclusive identification* with that personal sense of self that creates unnecessary boundaries and division to radical Wholeness.[41] This is

---

40. The work of Daniel P. Brown is most helpful here to point out how layers of confusion can be liberated. We follow his work closely for this vector. Dan Brown first learned some of these distinctions from Denma Locho.

41. Scientific findings from the field of cognitive neuroscience support the correlation between the increasing disentangling of awareness from thought and self with well-being. Farb et al. (2007) found that mindfulness practitioners (with mindfulness understood according to the Vipassana position of cultivated meta-awareness rather than the Tibetan definition as related to concentration) were shown to have uncoupled the usual connectivity between the regions of the brain that process immediate self-referential experience from those that support more narrative-based forms of self-reference (i.e. self-beliefs and self-construct born of life-experience over time), whereas non-mindfulness practitioners had not. Importantly, this uncoupling was correlated with higher reports of well-being. This supports the above statement that for the majority of human beings, awareness remains confused with thought, self, time, and individuation, but that through spiritual practice and the neuroplasticity of the

a point that has particular significance to the evolutionary awakening sub-line of identification.[42]

Even among those who are able to disentangle awareness from thought and their personal sense of self, many are still trapped in the confines of the coming and going of time/space construction (note the third concentric circle in the graphic above). Dividing things up into a past, present and future, as well as inside, outside, up, down, left and right, creates a differentiation that blocks recognition of the true source of awareness as the undivided sphere of the Totality. It is only by resting and surrendering into the ever-present moment of the eternal "now", without creating the separation of space by defining a world "out there", that Reality might be freed up to reveal itself more nakedly.

Finally, the most difficult hurdle to leap in the process of vantage point development is the association between awareness and individual consciousness (note the forth concentric circle in the graphic above). Inevitably, individual consciousness creates a sense of doership. In the advanced stages of the path, awakening is natural and effortlessly present, moment-to-moment. The Tibetans refer to this as non-meditation. As long as individual consciousness is still present, however, there is a concept that there is *something* to do and *someone* to be awakened. Individual consciousness perpetuates the thought that there is someone who will *gain* enlightenment. It is this final veil of separation that prevents the full, liberated recognition of one's true vantage point—the inseparability of ever-present Radical Wakefulness and the entire spectrum of form—Gross, Subtle, and Causal. Once individual consciousness is seen through as a wave on the ocean of Reality, Awareness has the potential to recognize its natural Source as the continually unfolding Base of Existence. Generally speaking, this is the Atman that is Brahman of Vedanta; the Parashiva of Kashmir Shaivism; the Suprepe Identity of Sufism; Dharmakaya of Buddhism and Bon. In this book, we follow Daniel P. Brown's lead and call this Ultimate and Primordial vantage point (which is no particular vantage point) *Awakened Awareness* (note the base upon which all the spheres are drawn).

In the process of radical awakening, when viewed from the relative perspective, each time awareness disentangles itself from a lower level of association it appears then to take a deeper basis as its vantage point.

---

brain, this confusion can be remedied, and the freedom and well-being of awakening stabilized.

42. In the context of evolutionary awakening a healthy individual self-sense is necessary for further kosmic unfolding.

Importantly, each deeper basis transcends, includes and penetrates the previous, shallower vantage point. This relates to the point made earlier in Chapter 4 on the shifts that occur as the proximal and distal sense of self, expressed along the path of radical awakening. For instance, the first vantage point is one that is exclusively identified with thought. Most of us (at least those of us in the industrial/modern world) find ourselves at this level of vantage point development. As such, right now, we can notice that naturally, without any effort, awareness as thought transcends, includes, and penetrates through the physical body. From this point of view, the physical body is a vehicle for awareness as thought to shine through.

The second vantage point is free from thought but still confined to one's personal sense of self. When awareness is thought to be a personal self-sense, it is obscured by wants, needs, desires, and aversions. It might also be confused with a role (mother, father, teacher, lover, etc.) or a societal affiliation (Christian, doctor, therapist, banker, etc.). This vantage point of a personal self-sense, even right this very instant, transcends, includes, and penetrates the physical body and thoughts. From this perspective, thoughts and the physical body are vehicles to express awareness as a personal self into the world—a useful function, indeed. This is the first vantage point to be self-reflexively established; consequently, Brown calls this second vantage point *Awareness*.

The third vantage point is free from a personal sense of self but still appears to be contained within the time/space continuum. At this level, awareness seems to come and go. It may also be spatialized as somewhere in the body or expansive and vast. This third vantage point transcends, includes, and penetrates the personality, thought, and the physical body, using all of them as vehicles of expression and manifestation. Brown calls this third vantage point *Awareness-itself*.

The fourth vantage point is liberated from its entanglement with time but is still associated with individual consciousness. At this level of realization, awareness is always present, effortlessly, in every single moment, but there is still a sense of an individual consciousness doing the practice. This vantage point correlates to the Vedantic state of the Witness—*turiya*. It transcends, includes, and penetrates through the coming and going of time, the personality, thought, and the physical body. Brown calls this fourth vantage point *Ever-Present Awareness*.

The deepest vantage point is one free from all obscuration, rightfully resting perfectly and stainless as the undivided Sphere of Reality. This vantage point is radically unbounded Wholeness. What is left, as a result, is a dynamic spaciousness of Awareness that naturally co-arises unified and undifferentiated from all appearances, all form, all movements of consciousness, thought and feeling—all depths of the entire kosmic spectrum of relative reality. From this Nondual perspective, Awareness brilliantly shines forth, transcending, including and penetrating all of existence as an expression (or as the Tibetan tradition says, as ornaments) of its own natural Wholeness.

When such a realization is made a stable recognition twenty-four hours a day and is allowed to penetrate through all layers of relative reality it is traditionally defined as enlightenment. It is important to note that true Nonduality transcends and includes duality as a relative expression of itself. In the terminology of the lineages of awakening, Nondual realization corresponds to turiyatita, sahaja samadhi, Nondual Suchness, and the Great Perfection of Dzogchen.[43] Brown calls this deepest vantage point *Awakened Awareness*. In the words of the great Dzogchen master, Longchenpa::

> There is no imperfection anywhere:
> Perfect in one, perfect in two, perfect in all,
> Life is blissfully easy.[44]

Other than a stable and continuous sense of self, there are no prerequisites for radical awakening.[45] The Absolute Reality of Awakened Awareness is available at all times and in all situations. It is the source and condition of all relative experience—the single State within which all states arise. No matter what an individual's spiritual development might look like in the other areas of growth, radical awakening to unbounded Wholeness is every human being's birthright. In this light, it is vital to remind the reader again that the stages described here represent a gradual path to awakening. Recognizing Ultimate Reality as Awakened Awareness does not *require* stages of opening, though paths that outline a gradual path through them are of deep service to many practitioners. Traditions like Dzogchen, Zen, or Advaita, for instance, emphasise sudden realization or the instantaneous capacity to cut-through all obscuration into immediate recognition of Reality.

---

43. The Tibetan tradition from which Daniel Brown teaches makes clear that there are several levels of nonduality. Only the final of which can be equated to Awakened Awareness. See Volume 1 of this series, *Streams of Wisdom,* for details.

44. Dowman, *Natural Perfection: Longchenpa's Radical Dzogchen*, 2010, p. 54.

45. Jack Englar is famous for saying that you have to have a stable sense of self before you can transcend it. We agree fully with Englar. If reality is deconstructed to its Absolute Base with out a stable sense of self/personality, the result can be devastating. A human being might suffer from a number of disintegrated pathologies or at worse, psychosis. A stable, healthy sense of self is the only requirement for radical awakening.

Radical Lines of Development

Here, we diverge from Brown and draw upon some of the elements of Wilber's work most applicable to this context. In any given moment, and in any given situation, awareness has the capacity to enact reality according to a 1st person, 2nd person, or 3rd person perspective. We find these same perspectives at the very foundations of our world's languages as well. A 1st person orientation refers to the person speaking, or "I". A 2nd person orientation refers to the person being spoken to, or "you". A 3rd person orientation refers to the person being spoken about, or "it". Wilber is known for establishing the primacy of perspectives over perceptions. This means that even before there is a perception of a subject and an object, there is awareness enacting that object as either a 1st person, 2nd person or 3rd person perspective. This has led to several interesting understandings with regard to spirituality and the radical awakening vector altogether.

The discovery most relevant to our discussion here is something that Wilber calls the Three Faces of God. In other words, because perspectives are primary, even God can be enacted from a 1st, 2nd or 3rd person perspective (in this sense God is understood as Causal Reality). God enacted as a 1st person is the Great "I" or True Self. God enacted as a 2nd person is the Great "Thou". God enacted as a 3rd person is understood to be the Great "It". Buddhism and Hinduism, for instance, in their practice-oriented and meditative lineages that focus on awakening to and as Buddha-Nature or the Absolute Self, tend to give ultimate preference to a 1st person perspective of the divine. The great Abrahamic traditions of Judaism, Christianity, and Islam, in their emphasis on relationship to God tend to give greater emphasis to the divine as 2nd person. Taoism, in its teaching on the Tao as Ultimate Reality, for instance, tends to give greater emphasis to a 3rd person perspective of the divine.

We bring this up here to point out that the path tracked at the beginning of this chapter using the work of Brown is but one example of a deep structural path within the vector of radical awakening as it is enacted from a 1st person perspective. By its very nature and due to the fact, at least in part, that it traces both Hindu and Buddhist paths, there is an explicit bias toward the Great "I". In the articulation above about vantage points, we tracked a 1st person perspective as it moved from Gross, to Subtle, to Causal, to Nondual.

It is important to understand for future projects that this is not the only way that awareness can travel back along the various states/phases of radical awakening. Just as we traced the stability of awareness as it moved through various phases of radical awakening through vantage points, the same sort of radical awakening can unfold through a 2nd or 3rd person perspective as well. Each of these perspectives can be thought of as a radical line of development. The graph below lists each of the perspectives as a particular line of development.

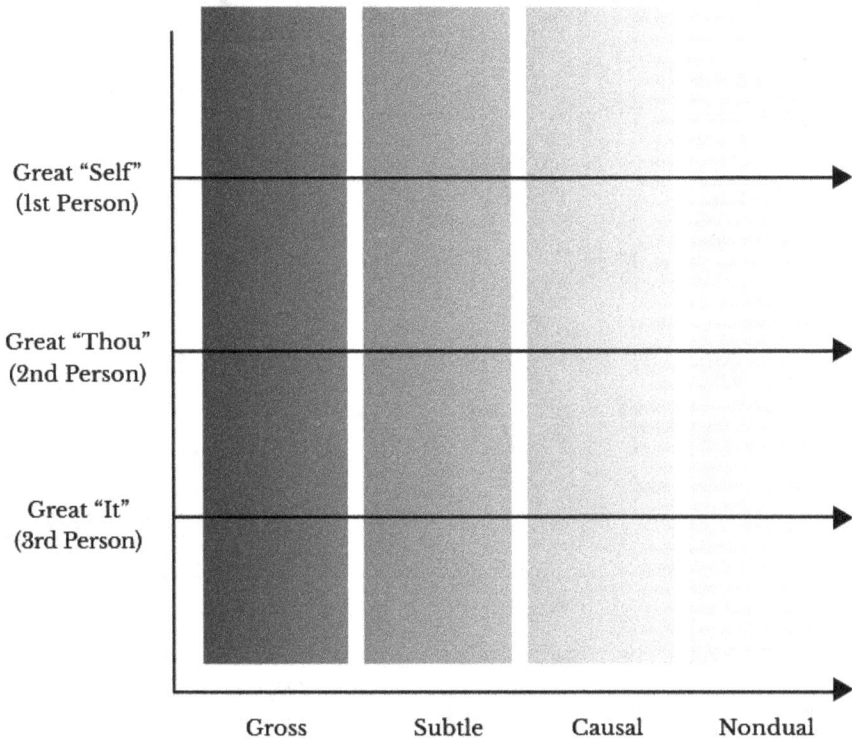

States/Phases of Radical Awakening

Figure 10. Chart showing the unfoldment of radical awakening through the states/phases of creation according to 1st, 2nd, or 3rd person perspectives.

It might also be the case that development does not unfold evenly within each individual (or each tradition). In this way, an individual might be fully developed through vantage points (state-stages of 1st person Awareness) but may not have a relationship with God (Causal 2nd person). Similarly, he or she may not have an understanding of the Infinite Field in which all is arising (Causal 3rd person). This person's radical awakening psychograph might read as follows:

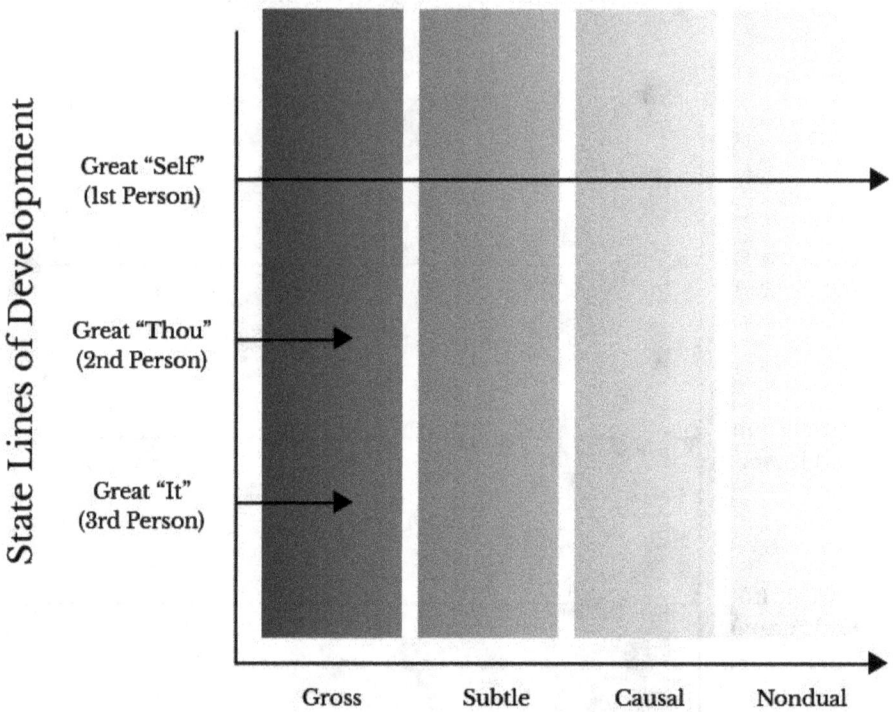

Figure 11. Chart showing how some individuals or traditions
may focus on a particular perspective—1st, 2nd, or 3rd—in radical
awakening.

Let us note that it seems that any of the three perspectives can be
taken all the way to Nondual Source. At Nondual Source, all perspectives
collapse into the single reality of the Unbounded Whole. I suspect this
will make certain room for 2nd person traditions such as Christianity
and Islam to hold the possibility of full Nondual Realization also.

All of this is to say that any future study interested in synthesizing a
truly integral, universal, trans-lineage spirituality will have to take these
sort of considerations into account if there is to be a true honoring of
the particularities of each of the traditions when it comes to radical
awakening.

PART 4: EVOLUTIONARY AWAKENING

In Being we are the one infinite and ever present reality. In Becoming we evolve eternally.

(Master Djwhal Khul and Bruce Lyon)[46]

---

46. Lyon, 2010, p. 234.

Earth is Eden

Chapter 8
The Nature of Evolutionary Awakening
Dustin

We use the terms planes and altitude interchangeably so as to create some coherence between the Trans-Himalayan system and Integral Theory. In the evolutionary vector we describe an identification line, a relationship line, and a line relating to plane access. Each of these, as we noted earlier, can be understood as self-related lines. We also describe a second group of lines (mostly cognitively based) relating to intelligences. Overall, the basic notion of development in this vector is fairly straightforward. Each of us, as relative selves, start development at the bottom of the rainbow spectrum and have the opportunity to proceed upward along the vertical axis as development unfolds. Notice the graphic below. At the bottom of the vertical spectrum you will notice the color infrared and next to it the word "physical". The color infrared refers to altitude in Integral Theory. The word physical refers to the plane in Trans-Himalayan cosmology.

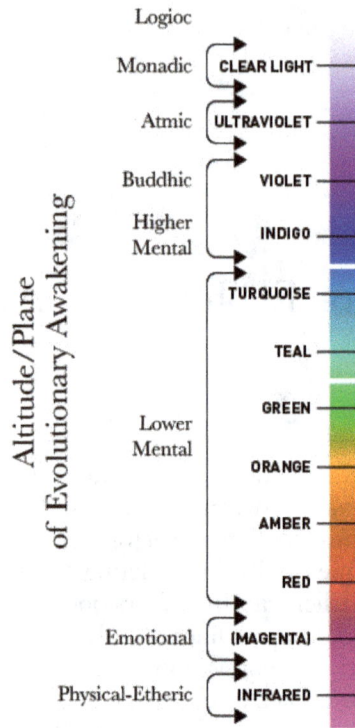

Figure 12. Graphic showing the correspondence between the spectrum of altitude used in Integral Theory and the planes described in the Trans-Himalayan teachings.

As each of us enters into the world as a human being, we begin in our mother's womb emerging into physical form. As development unfolds, we slowly progress up the spectrum of development. We emerge out of a strict reliance on the physical plane and move into the emotional plane. At this stage we have the most basic of feelings (safe, unsafe, pleasant, unpleasant, etc.). This plane is most closely related to a magenta altitude. At a certain point in the developmental progression, an individual may extend beyond the emotional plane and enters into the mental plane. The Trans-Himalayan mental plane contains within it several altitudes ranging from red to turquoise. In the subtlest levels of the mental plane we enter the altitude of the soul, which is normally understood to abide on the higher mental, buddhic and lower atmic planes. This corresponds to indigo, violet, and ultraviolet altitudes. As development continues to unfold even further, humans begin to reach into the monadic planes (higher atmic, monadic, and logoic). These levels of growth correspond to an altitude of Clear Light.

Even with these clear correlations between the notions set forth in the Trans-Himalayan teachings and those outlined in Integral Theory, there are still several distinctions and points of difference. Integral Theory understands each of these planes to be equivalent to levels of consciousness and not as separate realities in themselves. The Trans-Himalayan teachings, while agreeing with the point that planes have a relation to levels of consciousness, also see the planes as existing as ontological realities in their own right. For the Trans-Himalayan teachings, the planes are actually the bodies of greatly evolved devas, who exist whether there is consciousness operating on them or not. This is something we will return to in the next chapter. From this perspective, it is possible for beings to reside on a particular plane as its own ontological reality while not residing on others. This means that there might very well be beings operating through etheric or astral bodies that incarnate on and inhabit the etheric or astral planes. It means there might be beings operating exclusively through a mental body that inhabit the mental plane. There might be soul beings who are operating exclusively through atmic, buddhic and higher mental bodies on the atmic, buddhic and higher mental planes. And finally, there are beings operating exclusively through monadic bodies inhabiting the monadic planes. We explore all of this in much more detail in Chapter 11.

Evolutionary Lines of Development

In Chapter 6 we learned that there exist several lines through which radical awakening can open up according to various perspectives taken (realization of 1st person Absolute Self/Consciousness, 2nd person God/Goddess, or 3rd person Reality/Kosmos). The same is true here in the evolutionary awakening vector. In the next three chapters we explore three specific lines as they develop through multiple planes of reality. Specifically, along the line of a 1st person perspective (I), we track awareness as it moves through a process of deepening identification into wider and more inclusive spheres of identity through the evolutionary vector (Chapter 10). Along a second person perspective (you/we), we trace awareness as it opens to and cultivates relationship with various cultures of beings on multiple planes of reality (Chapter 11). And in terms of the third person perspective (its), we trace awareness as it penetrates into, and opens to being penetrated by, the energies of multiple planes of reality (Chapter 12). After exploring these first three lines, we explore a group of intelligence-based lines in Chapter 13 synthesized from Integral Theory. The graph below lists an identification line, a relationship line, and a plane access line of development.

Dual Center of Gravity

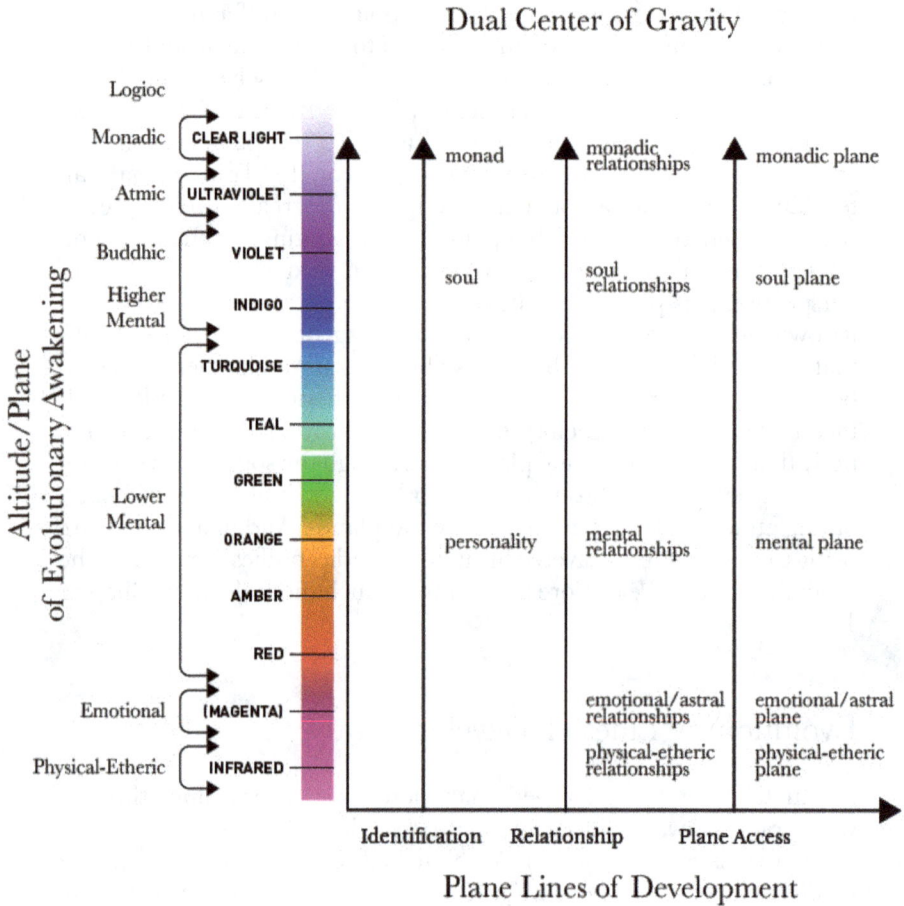

Figure 13. An evolutionary awakening psychograph showing lines of development extending through the planes.

A foundational part of our exploration in this book is that development in each of these lines can unfold unevenly. A person might have a very high level of development when it comes to his or her capacity to interact with beings on the buddhic and atmic planes, for instance, but may not have established a stable level of identification beyond the personality. Similarly, a person might have the capacity for relationships with beings on the atmic plane, but he or she may not have learned how to cultivate conscious access to the energy and information contained on such planes when those beings are not present. Both of these situations are often seen in people who do channeling work (when authentic). In fact, the person may not have stabilized access to navigate or operate freely on planes higher than the physical, as is the case with much of humanity. Such an individual might hold the following psychograph.

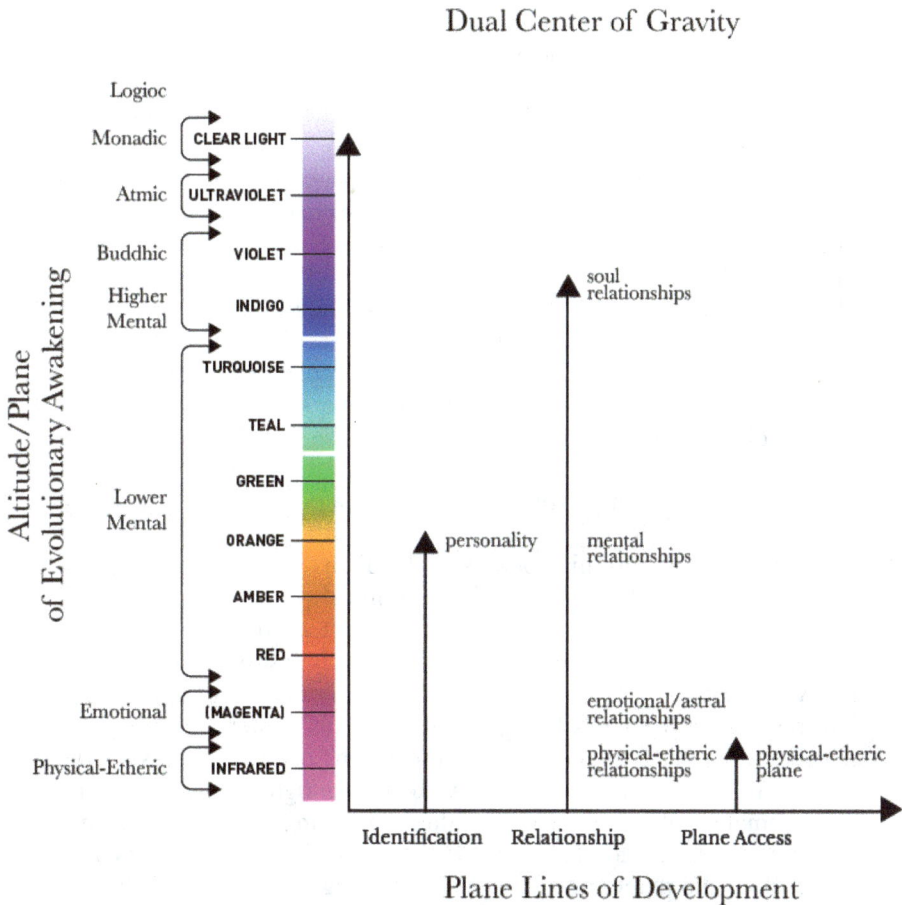

Figure 14. An evolutionary awakening psychograph showing lines of development extending in an uneven manner through the planes.

Evolutionary Awakening and Shadow

When shadow is present along the evolutionary awakening vector, spiritual practitioners may find themselves with aversions to particular altitudes of identification, relationship, plane access or intelligence unfoldment. Or even more damaging, if the pathology arises at the most foundational levels of self-development in the earliest stages of childhood, for instance, the individual may find that they cannot hold together a coherent and continual sense of self (e.g. borderline personality).

Integral psychology teaches us that all individuals have two aspects of self in relationship to shadow. The first is called the *persona*. The persona relates to the projected ideas and concepts a person has about who he or she is. The second aspect of self is what Wilber refers to as the *actual self.* This is the actual, true character of the self—the way a self *really* is. In most people there is a gap between the persona and the accurate self. This gap leads individuals to have an inaccurate experience of relative reality. In some cases, individuals create significant suffering in the world because they are not acting from a healthy, accurate sense of self. Western psychotherapy helps to close the gap.

All of this means that, like light shining through a dirty bottle, even if an individual is highly developed along the radical awakening vector, that realization still shines through a relative self with a particular degree of shadow in the evolutionary awakening vector. The less shadow, the more clean, clear and transparent the realization, and the more likelihood that it (if authentic) will have a positive influence in the world.

In the second and third phases of the Trans-Himalayan teachings, though not so much in the first Theosophical phase, there has been a healthy and strong emphasis on the re-integration of previously dissociated psychological material. As these teachings continue to evolve, it is vital that they continue to take these insights into consideration. No matter how awake a being might be in his or her vantage point realization (radical awakening), or the extent of their access to the heights in altitude (evolutionary awakening), right action and true transformation of the planet will be limited as long as the shadows and personas of beings cloud and distort the expression of the living Clear Light. Both the planet and kosmos need beings that are as healthy in their awakening (horizontally) and as clean in their alignment and identification with planetary purpose as possible (vertically).

Chapter 9
The Multidimensional Planes of the Kosmos
Jon

The planes can be understood as increasingly subtle levels of vibration along the evolutionary vector in which the Four Quadrants tetra-arise. They are gradations of energy density and information complexity, and can also be understood as the various dimensions through which Awakened Awareness emanates and reflects itself. In this chapter, we examine the planes through the LR quadrant as objective environments. These form the fields in which development along the radical and evolutionary awakening vectors occurs. In terms of their precedent in the spiritual traditions, the planes correspond to the various realms or increasingly pure abodes/pure lands that are found along the axis of Mount Sumeru in the Buddhist cosmology, the Hindu lokas or the various dimensions accessed by trained shamans, for instance.

Fundamentally, the LR collective energy fields, or planes, are the UR bodies of kosmic beings incarnating at higher octaves, which themselves can then be recognized as holons situated within an even higher octave of LR collective energy fields, which are again the UR bodies of *even larger* kosmic beings, ad infinitum....

I know that seems complicated! So let's break it down a bit. The UR buddhic body of a human being exists within the LR buddhic energy-field (or plane) of the Earth. This LR buddhic energy-field of the Earth is a *part* of the UR etheric body of the planetary Logos of the Earth. The UR etheric body of the planetary Logos of the Earth exists as a part of and within the LR etheric energy field of the solar system. The LR etheric energy field of the solar system is the UR etheric body of the solar Logos, and on it goes...

A Post-Metaphysical Consideration of the Planes

The planes are definitely one aspect of the pre-modern esoteric cosmologies that will need to be updated according first to modern scientific and post-modern perspectival requirements, and then integral requirements, in order to be part of a post-metaphysical universal spirituality. Ultimately from an integral perspective, we find it necessary to include the recognition that the planes are not ontologically pre-given, but are actually evolutionary grooves that are laid down over time. With this understanding, we also see planes as dimensions of reality that are continually enacted by individuals and collectives at particular frequencies and scales of the kosmos. The process of enactment includes the influence of all Four Quadrants (consciousness, body, culture, collective systems). It is only through the confluence of these factors that we have any planes to speak about.

However, before we are able to arrive at such a sophisticated understanding of the planes, it may be helpful first to consider them in terms of the requirements of modern science and then post-modern constructivism. Once these more conventional approaches to epistemolgy have been integrated, we can then return to a more integrally-informed, multidimensionally-oriented consideration.

Modern science lays particular importance on the disclosure of 3rd person evidence from any particular experimental inquiry. Therefore, one way to begin to understand the planes that integrates the requirements of modern science is to consider them in the light of how we understand perception.

First consider that human beings are only able to see a very small range of the total electromagnetic spectrum (390-700 nanometers). To put that in perspective, if the entire electromagnetic spectrum was represented in terms of the length of a reel of video film that covered a 2000-mile distance, the portion of it that human beings can physically see would comprise only 1 inch. That leaves a vast amount of 'stuff' going on the kosmos outside of the bounds of our capacity for perception—the vast majority, in fact! Indeed, this illustration of the limitations of human perception to fully know the 'stuff' of which our universe is made is supported by the finding that only 4% of the mass of the universe is accounted for by known matter. 96% remains unaccounted for, and is comprised, according to current theories, by dark matter and dark energy.

From the exterior perspective, what human beings see as the visible 'outside' world and its various objects are essentially patterns of radiant light and energy that exist in this visible range of the electromagnetic spectrum. Modern science tells us that we do not perceive things the way we do because that is how they actually are, but rather owing to the biological makeup we have. Of course, as the post-modern theorists have called us to recognize, it is not just our biological makeup that shapes our perception, but also the subjective, cultural and socio-historical perspectives we inhabit also. For now though, prior to entering a consideration of the post-modern demands in terms of how we understand the planes (which actually radically shifts the topic of our discussion *from perception to perspective*), let's stay with the current exploration from a conventionally scientific orientation, just to be able to meet its particular requirements.

To return to the presentation given to us by modern science, we see the visible range of colors we do because in the retinas of our eyes we humans have two kinds of photoreceptor cells—cones and rods. These detect certain visible wavelengths of the electromagnetic spectrum and translate them into neuroelectrical signals that are then relayed to the visual cortex, and then to the visual association cortex in the brain, where they are converted into the recognized representation of a particular object. Make no mistake, what we see is a *representation* of the actual object or set of objects as they exist only in the UL consciousness, the UR brain and the LL culture of the perceiver. We do not see the object itself. That which is seen, and which is translated into a perceived object or set of objects, is formed from bandwidths of radiant light-energy that exist in the visible region of the electromagnetic spectrum. As Immanuel Kant pointed out over 200 years ago, we can only ever know the phenomenon, not the noumenon. That is, all we can ever know is our experience of a thing; we can never know the thing in itself. That is unless, as the radical awakening lineages tell us, we go deeper than the first order subject-object dualism that apparently separates perceiver and perceived in the first place.

The light and energy that composes the visible region of the electromagnetic spectrum is what we collectively see as the physical world. The fact of relative consensus in perception—that the vast majority of us seem to see the same things when we look 'out' at the world (a tree with the same structure of branches, for example)—is not evidence for us being able to truly see the world as it is, but is rather (from the UR perspective) a product of the vast majority of us sharing essentially the same neurobiological 'glasses' through which we perceive it. In this, we have a modern, scientifically-informed presentation of the fundamental Buddhist teaching on Emptiness (*sunyata*). The doctrine of Emptiness teaches us that all we conventionally know of ourselves and the world are simply constructions—constructions of thought, feeling, and perception—which do not necessarily represent the world as it *truly* is. It is a razor-edged teaching that critically deconstructs the way human beings conventionally conceive of the entirety of relative reality, leaving only a giant question mark remaining. That question mark is an injunction towards practice. It is an injunction towards radical awakening. If you want to know things as they truly are, you must recognize that which is prior to their arising, prior to your arising, prior to the arising of culture, society and time itself. You must awaken as the Clear Light that is the basis of all.

To return to our consideration of the planes according to the position of modern science, the rest of the electromagnetic spectrum beyond this 390-700 nanometers that is viewable by the human eye is composed of wavelengths of light-energy which are not conventionally perceivable to the human biological apparatus at present (though animals such as birds and insects such as bees, for instance, are able to see outside of the visible band as it exists for humans). However, that does not by any means indicate that those wavelengths of light-energy might not also comprise patterns of organization that collectively form what we might interpret as worlds (in company, that is, with the types of subjectivity, body, and culture that enact them). This is how we would begin to point to what in the esoteric and shamanic lineages are called planes according to a post-metaphysical approach.

A crucial component to the post-metaphysical grounding of the teaching on the planes is the recognition that their existence can be supported by evidence gathered according to the injunctions of the scientific method. Drawing on and expanding William James' presentation of radical empiricism, Wilber has made the point that the fruits of spiritual practice are verifiable according to what he calls "the three strands of deep science". These are, (i) conducting an experiment, (ii) collecting the data after the prescribed period, and finally (iii) cross checking it among a community of peers who have also done the same experiment to compare your findings.[47] As we look to some of the esoteric and shamanic lineages and pathways that have articulated the existence of there being multiple planes of existence beyond the physical, what we find is a detailed literature that presents specific injunctions in terms of practice, which may be worked with to open up such access. What we discover when we study those texts and engage in those practices is that the recognition of the kosmos as having multiple dimensions of consciousness and energy-matter to it is something that is scientifically reliable. That is, it is a consistently reproducible phenomenon for those who develop the requisite capacities, with the the capacity to do so being definitely possible of cultivation if the prescribed practices are engaged in correctly.

Now, though there may be evidence presented for the existence of multidimensional planes of experience through the finding that experience of them can be opened up to by anyone who employs the

---

47. Wilber, 1998.

right approach (deliberately or not), the next question any scientist looking at such a situation would be interested in is that of validity. While *reliability* points to whether a particular result from a scientific experiment is consistently reproducible, *validity* points to whether it is actually representative of reality. To explain this, consider the following example. You and I might both engage a practice where we lightly apply pressure on our closed eyes with our fingers, which produces for us both the appearance of particular patterns of light to our sight. But does the fact that after we both put such pressure on our eyes we start to see these patterns mean that we are suddenly able to see more of reality than we were before? It more likely just means that both of us have done something to produce particular common effects in both of our perceptual faculties. The fact that you and I get the same result again and again when we repeatedly apply such pressure to our closed eyelids satisfies the scientific requirement for reliability. Validity, though, is normally only satisfied by checking our findings back against already-established theory and evidence. In the current discussion of planes of the multidimensional kosmos, this would involve checking our experience against the way in which the esoteric and shamanic lineages outline their practice-findings. Of course as time goes on, and as broader and broader sections of humanity continue to engage with such explorations so that the boundaries between different spiritual communities get thinner, it is likely that the theory and evidence base that we can draw upon to consider such questions as this will continue to grow outside of the esoteric and shamanic lineages. This growing theory and evidence base will be crucial in grounding the understanding of the multidimensional planes or realms of the kosmos in a post-metaphysical universal spirituality.

Since the question of validity deals with the nature of reality, it is necessarily one that has philosophical and epistemological significance. This is where we begin to address the post-modern requirements for any mature and post-metaphysical consideration of the planes. As Wilber has noted, the central recognition of the great post-modern theorists (e.g. Derrida, Foucault, Wittgenstein) is that whatever perceptions we might have, they are always situated in inter-subjective cultural contexts that influence or constrain them. This is where we recognize that rather than speaking of perception, according to the post-modern theorists, it is more appropriate to speak of perspective.

As an example of this, let us consider two Tibetan monks who are advanced meditators in concentration practice. The Tibetan meditation texts are incredibly detailed when it comes to describing and defining multiple forms of meditative experience, including everything that pertains to concentration practices. One particular teaching on this comes from the great Buddhist sage, Asanga, who is understood to have received it from the Buddha to come, Maitreya, on the subtler planes of existence. This teaching describes in fine detail nine stages of increasing-until-fully-stable concentration (*shamatha*). At a particular point of these nine stages, there is such a stabilization of concentrative absorption that a practitioner's perception of all physical plane phenomena collapses and their awareness transitions onto subtler planes of experience. Incidentally, this should not be understood as the end point of concentration practice, but rather a natural by-product of one of its more advanced stages of accomplishment.

Now, returning to our example of the two meditating Tibetan monks, let us imagine that through their advanced concentrative training they both experience this transition of awareness onto subtler planes. Let us further imagine that they begin to perceive some of the Subtle state phenomena that pervades those planes, such as deity forms and subtle mandalas.[48] These two monks perceive and recognize these deity forms as particular Buddhist deities or celestial bodhisattvas, such as Avalokiteshvara or Vajrapani, for instance, clothed in traditional garb and with Tibetan facial features. This is where the post-modern theorists come in to raise the question: "Haven't you ever noticed how all those depictions of deities, of angels, of gods, goddesses, and other celestial images are always represented with the same cultural and ethnic features as the community of people living in that particular place? Isn't that kind of a strange coincidence? Tibetan Buddhist deities have Tibetan ethnic features. Christian angels are depicted in European history with white European features. Ancient Mayan deities were represented with ethnic Mayan features. Strange, huh?" The post-modernists have raised a challenging point here.

What could be the possible answer to this question? Does it indicate that the same ethnic diversity that exists among humanity exists among subtle phenomena, so that we should really speak of Christian angels being white European, or Indian deities being Indian? Or could it mean that what is really perceived by such folks as our two above-mentioned Buddhist monks, along with every other being of every culture through time who has had such experiences, is just a product of their own mind?

---

48. It is important to note that Wilber's Integral Theory sees these as subtle state-phase manifestations (i.e. states) and not as plane manifestations (i.e. structures). In our model, we make the differentiation between the state-phases of Reality's emanation of the kosmos into form (from Causal to Subtle to Gross), and the multiple dimensions, or planes, of the evolving kosmos once emanated. We further suggest that deeper and deeper state/phases (Gross to Subtle to Causal to Nondual) are most prominent to consciousness on higher and higher planes. The rationale for all this is something we go on to unpack more fully later in this chapter.

An Integral approach to answering this question takes the middle way between both of these extreme positions. It would look to integrate all the perspectives raised within a multi-perspectival and developmental context. It would suggest that none of the above needs to undermine the point that those spiritual practitioners who have reported such experiences as described here have genuinely shifted their awareness onto deeper states/phases of creation, or subtler planes of experience, than the physical. An Integral approach would just additionally recognize that once they have, their experience of whatever phenomena arise is colored by the subjective, biological, cultural and socio-historical assumptions and contexts within which such individual experiences occur. Furthermore, it would recognize that all phenomena are enacted by the developmental level of the individuals and collectives in question. This is the integration of perspectives (the Four Quadrants) and developmental levels (states and structures) that characterizes an Integral approach.

In order for any presentation of the planes to be able to show up fully at the post-metaphysical round table of a universal spirituality, it will necessarily need to satisfy all these criteria. It would need to ground that presentation in empirical evidence that tallies with modern science. Additionally, it will need to inculcate recognition of the subjective, biological, cultural, and socio-historical relativity of perspective and perception across all planes of experience described by the esoteric and shamanic traditions. And of course, it will need to do this within a developmental context.

This calls us to recognize that on what we will go on to describe as the soul planes, for instance, a soul's perception of the shared environment, which exists in a non-physically visible bandwidth of the electromagnetic spectrum for incarnated human beings, will necessarily be constrained by the perspectives inherent in its Four Quadrants. That is, the experience will necessarily be influenced by the particular idiosyncrasies of its UL consciousness (that individual soul's particular karmic history, values and perspectives). It will be colored by the composition and anatomy of its UR soul-energy-body (perception and contact with its soul-level environment here will be dependent on the subtle anatomy of the soul-level perceptual system through which a soul is operating). It will be shaped by the perspectives inherent in its shared culture in the LL (the collective community of souls with whom it shares inter-subjective relationship), and also by the type of energy that predominates in that LR field of energy-matter (which planes it is operating on). This would be the same on what we will go on to describe on the monadic planes also, though in a very different way, owing to the monadic planes being planes on which the illusion of duality has been seen through in the ever-present realization of Nondual Divinity.

Taking such an approach as this begins to obviate the errors inherent in what Wilber has called the "Myth of the Given".[49] The Myth of the Given is the assumption, common to the pre-modern spiritual traditions, that reality is an independently existing thing "out there", changeless across observers. This relates to the degree to which we understand the planes to be pre-given ontological realities that are just simply 'there', like the pre-modern traditions that articulated the Great Chain of Being did. As Wilber has made the point, there is no way that such an assumption can hold in the arena of a post-metaphysical universal spirituality. Interestingly, the Trans-Himalayan teachings do not hold that the planes are ontologically pre-given, with them and their contents simply existing prior to and untouched by evolution. They do, however, stretch the modern and post-modern mind in their description of how this is so.

Wilber has challenged the pre-modern understanding of the planes, what he calls structures, as pre-given, by first making the important differentiation between states and structures (what we describe as radical and evolutionary awakening), so that states are not simply stacked on the top of the evolutionary structures. He has also posited that rather than the structures being pre-given, they are actually evolutionary emergents that unfold through the pioneering activity of human beings who break through into higher, new structures of operation. How, then, can the Trans-Himalayan teachings account for sets of planes extending deeper and higher into the kosmos, way beyond the leading edge of where human evolution currently stands?

The answer to this is that the Trans-Himalayan teachings, in contrast to many advocates of Integral Theory, *do not* see human beings as the leading edge of kosmic evolution. Rather, they understand humanity as one evolving community of lives within the kosmos among many others, each with its own areas of particularly profound development, but not the leading edge of all. The Trans-Himalayan teachings understand humanity on Earth to be one step on the evolutionary ladder here, beyond which there are many more, populated by beings of similar and post-human levels of development (masters, buddhas, devas, etc.). As such, the teachings describe that the opportunity to grow beyond that which we conventionally understand as human is embedded into the kosmic algorithms, or grooves, of evolution. These are the habits of the kosmos that have been carved out, like water cutting through rock so as to create streams, rivers and then canyons, by those in previous ages who have blazed the trail of which nirvana is the beginning.[50]

---

49. Wilber, 2007.

50. An important caveat to include here is that we are not suggesting that the most

To set the stage for our discussion of the sub-lines of the evolutionary awakening vector—identification, relationship, plane access, and various intelligences—we include below a tabulation. This shows what are understood as the seven sub-vibrational frequencies of energy-matter, or sub-planes, of the kosmic physical plane in the Trans-Himalayan teaching. And it includes the levels of identification, culture, energy, and intelligence, which are understood to be specific to those planes. In this, we maintain that access to these planes of being is a scientifically reproducible phenomenon for anyone who does the requisite practice. We maintain a post-modern sensitivity to the fact that the way the Trans-Himalayan masters have described and presented these planes is necessarily relative to their own level of consciousness; to the vibrational frequency of the soul and monadic-level bodyminds through which they are operating; to the shared culture of the Himalayan Branch of Hierarchy that they together compose, including the relative nature of the names and linguistic terminology used to describe the planes; and the types and forms of energy that pervade and flow through the collective matrices that are the planes on which they operate.

---

sophisticated mental plane *intelligences* (modern, post-modern, integral) were fully unfolded by advanced beings of previous times. Rather, our model posits that in order for a being to transition in their evolutionary altitude of identification from person-ality to soul, they need only have unfolded the highest level of intelligence available at their time of incarnation. Therefore, while we are suggesting that in previous ages human beings have developed into higher altitudes of identification, relationship, and plane access than humanity currently recognizes, we still retain the suggestion of Integral Theory that the highest structures of intelligence that are available now are as high as they have ever been. We just contextualise and situate that claim as relating to the evolutionary awakening sub-line of intelligence cultivation, specifically.

Integral Altitude	Level of Identification	Culture	Plane Energy	Intelligence	Plane	Radical State of the Absolute most pervasive
Clear Light	Monadic bodymind	Shamaballa	Planetary Life, Destiny, Will, Power	Supermind	Logoic	Nondual
					Monadic	
					Higher atmic	Causal / Witness
Ultraviolet				Overmind	Lower atmic	Subtle
Violet	Soul bodymind	Hierarchy	Pure Love-Wisdom	Intuitive mind	Buddhic	
Indigo				Illumined mind	Higher mental	
Red – Turquoise	Personality bodymind	Humanity	Intelligence Creativity	Egocentric – vision-logic	Lower mental	Gross
Magenta	Emotional self	Animal Kingdom	Emotional energy		Astral	
Infrared	Etheric impulsive self	Plant Kingdom	Etheric energy, Chi / Prana / sexual-energy		Higher physical: etheric	
	Physical self	Mineral Kingdom	Kundalini		Lower physical: dense	

Table 1. A table showing the correspondences between Integral Theory and Trans-Himalayan taxonomies of the levels of altitude, identity, community, plane energy, and intelligence, as they extend through the evolutionary planes.

The Relation between Radical States and Evolutionary Planes

The matter of how we understand the relationship between the various states of Absolute Reality (Gross, Subtle, Causal, Nondual) and the evolutionary planes is profound to consider. Naturally, both Dustin and I came to this discussion with the perspectives of the lineages we come from; Dustin bringing Wilber's clear distinction between states and structures, and myself with the distinction between radical and evolutionary awakening. These insights from our respective backgrounds led us eventually into some important clarifications. We recognized that the lack of differentiation between radical awakening and evolutionary awakening (or growth tracked through state-stages and structure-stages) has been a major source of confusion in spiritual teaching in the past. In his earlier work, Wilber made the error of conflating states with structures, and in the first two phases of the Trans-Himalayan teachings the differentiation is seldom clearly articulated. In fact, many teachings, subtly or not-so-subtly, assume that the states of the Absolute as described in Integral Theory, whose roots go back through to the Vedantic teaching and that of Tibetan Buddhism, should be equated with the evolutionary planes (structures in Integral Theory). Although there are some similarities between certain aspects of the planes and certain elements of states, the consequences of such a position severely compromise the integrity of that which we understand by both.

There are a number of intractable implications to a position that conflates planes with states. Specifically, when you do not differentiate between them and suggest that the Gross, Subtle, Causal and Nondual states embody an alternative terminology for whatever number of evolutionary planes another system of cosmology might posit, then:

a) The planes have to stop once they reach Nondual, thus screwing a very firm lid onto the scale of our worldview. For instance, once a being uncovers and stabilizes Nondual realization, there is then *nowhere else to go* and the journey is understood to have finished. Where does that leave both the Ancient Egyptian and Trans-Himalayan teachings, for instance, that describe paths of kosmic unfoldment that come online in post-human stages of mastery?

b) The Nondual Absolute, which is pointed to by the deepest radical awakening traditions as the highest realization, becomes implicitly something that is 'up there', only able to be reached on the 'highest planes' and in the highest levels of development.

c) Matter, or the form aspect of Absolute Reality, is cut off from its infinite root and made finite, limited, and something whose spectrum stops once we get to the heights. This consequently causes a deep schism between the Awareness and Form aspects of the Absolute in which we then have to understand only the former to be infinitely unbounded.

This will not do! Happily though, the model that we use in this book is able to address each of these questions in a manner that maintains the integrity of the Ultimate Sphere of Reality and its kosmic evolutionary expression. To demonstrate this, let us address each one of these intractable positions in light of the model we offer in this book.

In relation to a) we would point out that one can be radically awake and continue evolving indefinitely, from the personality to soul to monadic bodyminds and beyond. As we explore later, it is likely the case that radical awakening becomes a prerequisite for most advanced stages of development and mastery on Earth, and in the kosmos.

In relation to point b) we posit that, just as the radical awakening lineages have always taught, radical awakening is available to beings on every plane as long as they have developed identification to the level of having the capacity for conscious self-reflection (lower and higher mental in Trans-Himalayan language). This points to the Tibetan understanding of the preciousness of human birth as the capacity for conscious self-reflection required for Awakened Awareness to be able to recognize itself in radical awakening is only available to human beings. We honor this teaching, and yet we also add to it emerging neuroscientific research on the most intelligent animal species—such as cetaceans, elephants and certain primates, for instance—that have all been shown to have the capacity for self-reflective cognition. We would thus suggest that on this basis, in theory, such beings as these would also have the capacity for radical awakening to Ultimate Reality. And we would furthermore point out that while radical awakening is available to all beings on every plane as long as they have developed the capacity for self-reflection, radical awakening becomes progressively naturally evident as the relative self ascends through and into subtler and subtler planes. This is a point we will return to later.

In relation to point c) we would point out that true radical awakening necessarily shines through all planes of reality. Therefore there is the possibility of complete integration of Awareness and Form, always.

The Relation between Radical States and Evolutionary Planes in Detail

In his more advanced work on integral spirituality,[51] Ken Wilber has suggested that the structures of evolutionary development (1st, 2nd, and 3rd tier) each partake in a particular state, or vantage point. For him, the 1st and 2nd tier structures (magenta all the way through to turquoise) all involve awareness situated in the Gross state, but as 3rd tier structures unfold one gains access to evolutionary altitudes that begin to objectify the Gross state of the kosmos, then the Subtle state, and then the Causal state. Ultimately, this means that as the higher reaches of evolutionary structures unfold, deeper vantage points necessarily come online. As Wilber sometimes explains it, the actualization of *Supermind* (one of the higher stages along the evolutionary vector)[52] requires the

---

51. See Ken Wilber's *Integral Spirituality: A Deeper Cut.* Audio.

52. Wilber considers supermind the "highest" stage at the leading edge of the evolutionary vector. The Trans-Himalayan teachings consider it a "higher" stage (rather than highest), owing to the fact that there are stages of evolutionary development, and

recognition of Nondual Awakened Awareness, but the recognition of Nondual Awakened Awareness does not require the actualization of *Supermind*.

Bruce Lyon and Djwhal Khul present a similar approach in their work,[53] and we find both of these approaches in perfect harmony with how the different altitudes of identification, culture, and plane in the Trans-Himalayan teachings each partake in deeper states, or vantage points, of creation. According to our understanding, as represented in Figure 15, on the most subtle three planes (the logoic, monadic and higher atmic) of the kosmic physical plane, the Nondual and Causal states of the Absolute Reality are most nakedly exposed. Thus, the community of beings that make up Shamballa, which we shall go on to describe in Chapter 11, abide with vantage points anchored in the glory of the Nondual Great Perfection and the Causal unmanifest void. In terms of the Causal state's pervasion of the monadic planes, these are the levels are that the Master Morya refers to as those of "the Lion of the Desert".

---

kosmic octaves, beyond it. We address this questin more fully in Chapter 17.

53. Lyon, 2010, p. 56.

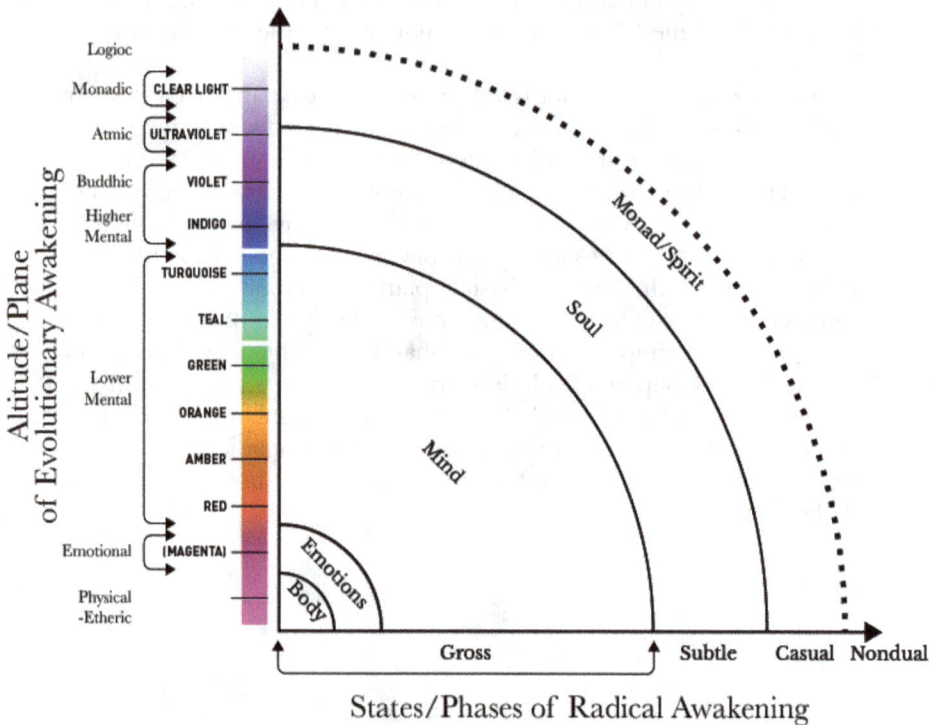

Figure 15. Chart showing the interpenetration of radical and evolutionary vectors resulting in the arising of progressive levels of identification.

On the soul planes (the lower atmic, buddhic and higher mental), the Subtle state of Absolute Reality is most pervasive, making these planes pervaded by forms of variously colored light, sound and subtle energy in which soul body-minds are anchored in Subtle vantage points.

And on the three densest planes (the lower mental, astral, and physical-etheric) the Gross state of the Absolute Reality is most prominent to our vision. Here, personality body-minds of individuals tend to be anchored in Gross vantage points, hence the more materialistic orientation of perception of the majority human beings, whose center of self-gravity resides on these planes.

These points have some interesting implications in terms of the path of awakening and evolution. They point to the fact that the probability of radical awakening increases as evolutionary development ascends vertically along the evolutionary vector, until it becomes effortlessly present in the monadic altitudes. This is precisely because the layers of horizontal state/phase obscuration are thinned through increased vantage point anchors as vertical evolutionary growth unfolds. It may even be the case, as Wilber speculates, that the continuity of identity into higher planes of evolutionary unfolding cannot be entered into without radical awakening.

The Seven Sub-Planes of the Kosmic Physical Plane

Having considered the radical context within which we understand the evolutionary planes, I will now describe each of the seven planes as they are presented in the Trans-Himalayan cosmology through an Integral lens. The idea here is that *each plane can be enacted through all four quadrants*, and as I go through each plane I will describe how each of those four perspectives shows up. It is important to remember also in this connection that these four perspectives relate to the sub-lines of our evolutionary awakening vector. The UL and UR relate to identification. The LL relates to relationships. And the LR relates to plane access. The line of intelligence can be considered as included in the UL.

Lastly, it should be noted that as we will explore further on, these seven planes embody *the seven sub-planes of the kosmic physical plane*, and that transcending, including and interpenetrating these there are, according to the Trans-Himalayan teachings, further kosmic planes (the kosmic astral, kosmic mental, kosmic buddhic, kosmic atmic, kosmic monadic, kosmic logoic, and even beyond these) to which access may be gained in post-mastery stages of unfoldment. It is upon these planes that radically awake beings of extraordinary levels of evolutionary development are understood to be incarnating. Here is the breakdown of the seven sub-planes of the kosmic physical plane from the subtlest to the most materially dense:

The logoic plane (clear light +): Beings whose altitude of consciousness inhabits the logoic plane would be described as buddhas in the Trans-Himalayan teachings. Here, in the UL quadrant of this plane, such a buddha's consciousness is characterized by natural radical awakening to the Awakened Awareness that is arising as and expressing through and as their monadic altitude of consciousness. This monadic consciousness is identified with the planetary Logos of the Earth—the vastly evolved kosmic Life that is incarnating through our planet and that I will go on to describe further later. Or it may be identified with other extra-planetary Logoi who are incarnating through planetary, solar, constellational or galactic fields. The Trans-Himalayan teachings use the term Logoi as a plural form of Logos. The Logoi are the radically awake and vastly evolved kosmic Lives who having awakened to buddhahood many ages ago are now incarnating through such massive structures as planets, solar systems, contsellations and galaxies, to hold fields for the awakening and evolution of all lives therein.

In the UR, having one's center of gravity on this plane enacts the subtlest layer of the monadic body. Through this subtlest monadic layer, the planetary Logos of the Earth's Will-to-Be and to express fully the entirety of its kosmically enlightened consciousness and being through its global physical incarnation is transmitted. In the LL dimension of this plane we find a sub-community within the collective culture of Shamballa that is oriented towards alignment with extra-planetary sources of consciousness. And in the LR of this plane, we find the collective energy-field of Shamballa that is constantly receiving the in-pouring shakti of extra-planetary energies that are contributing to the awakening and evolution of all life on Earth.

The monadic plane (clear light): In the UL quadrant of this plane we have the mid-level of the monadic domain. Here, a buddha's consciousness is again characterized by radical awakening to Awakened Awareness plus monadic identification. The UR quadrant again involves polarisation in the monadic body, which on this plane transmits our planetary Purpose as it is envisioned in the heart-mind of the planetary Logos of the Earth. The LL quadrant contains the collective culture of Shamballa—one radically awake consciousness that is in alignment with that of the planetary Logos of the Earth and various extra-planetary Logoi that are incarnating through planets, solar systems, constellations, who share relationship with our planetary being. And the LR quadrant contains the collective field of Shamballa that is a reservoir of the awesomely dynamic energy of our planetary Purpose.

The atmic plane (ultraviolet): As is shown in the tabulation above, the Trans-Himalayan teachings divide the atmic, mental and physical planes into two levels—the higher, subtler level, and the lower, dense-material level. On the atmic plane as a whole, the Trans-Himalayan teachings speak of masters rather than buddhas, and the UL quadrant of the higher atmic plane is enacted through a master's natural opening to the Causal state and identification in the grossest layer of monadic consciousness. The UR quadrant of the higher atmic plane involves abidance in the grossest layer of the monadic body, and is the level where the transmission of our planetary Purpose is translated into the energy of pure evolutionary Will that drives what the Trans-Himalayan teachings describe as 'the Plan'. While the planetary Purpose transmitted by Shamballa is understood as the alpha and omega point for the entire planetary manifestation of Earth, the Plan, which is held by and implemented by Hierarchy, is understood as however much of the Purpose can be worked out in a particular time cycle. One example of such a cycle is the presently dawning Aquarian

Age of approximately 2,300 years, or the greater solar year of 26,000 years, for instance. This is the level upon which Hierarchy, the planetary meta-sangha that we shall go on to describe in Chapter 11, receptively opens to the inpouring of streams of planetary Purpose and Will from Shamballa so as to formulate the Plan. The Plan is then expressed as the driving movement toward planetary service along a multitude of lines, which lives in the core of every soul. In the LL quadrant, the higher atmic plane is enacted by the sub-community of beings within Hierarchy whose role it is to hold a bridge in consciousness between Hierarchy and Shamballa. And in the LR, the higher atmic plane is enacted in the collective field of atmic transmission where the energy of planetary Purpose is formulated into a reservoir of evolutionary Will-force.

The UL of the lower atmic plane is enacted at the most subtle level of the soul. This most subtle level of the soul could be described in terms of a master's consciousness being identified with the generative spiritual Will of the planetary Logos of the Earth to express fully its divine nature through its planetary manifestation. In the UR quadrant of this lower atmic plane, the energy of this planetary Will is transmitted via a master's atmic body into the field of love, consciousness, and energy that is Hierarchy. In the LL, the lower atmic plane is enacted by a community of beings within Hierarchy who together participate in the full consciousness of the Plan and transmit that into the buddhic and higher mental levels of Hierarchy. In the LR of the lower atmic plane, we have the collective field held by these beings. This is a reservoir of pure evolutionary Will that pours into and orients the buddhic and higher mental fields of consciousness and energy within Hierarchy.

The buddhic plane (violet): In the UL, this plane is enacted through the intuitive consciousness of the soul that is deeper than mind, which spans across incarnations and karma, and contains natural capacities for telepathy, clairaudience, clairvoyance, and other siddhis. Here is the level of cognition that Wilber describes as intuitive mind. In the UR, this consciousness expresses through the buddhic sheath that is composed of pure love energy. In the LL, this plane is enacted through the collective culture of Hierarchy, the field of all awakened souls on Earth, and is simultaneously in contact with various other communities of souls incarnating throughout our galaxy. In the LR, we have the collective energy matrix of Hierarchy in which all souls move, which is a field of pure love and wisdom.

The mental plane (higher = indigo; lower = red-turquoise): The UL of the higher mental plane is enacted through the mind aspect of the soul's consciousness. These higher levels of the mental plane are where we find what Wilber calls illumined mind. The UR of the higher mental plane is enacted through the mental layer of the soul body, which is described in the Trans-Himalayan teachings as having an energetic form similar to a lotus. In the LL, we have the collective culture of all awakening souls among humanity, and in the LR, we have the collective energy field of soul-love in which they live, move and have their being.

The lower mental plane is enacted through the operation of conventional intellect and cognitive mind. It is the level of concepts, thoughts, and ideas. The UR of the lower mental plane is enacted through what the Trans-Himalayan teachings describe as the lower mental body, which expresses through the neo-cortical layer of the brain. The LL is inhabited by all intellectually-active humanity. And the LR is enacted as what Tielhard de Chardin called the noosphere— the collective matrix of all thought on Earth. It is on this lower half of the mental plane that we suggest the memetic structures of intelligence described in Spiral Dynamics and Integral Theory from red through to turquoise develop.

The astral plane (magenta): The UL consciousness of the astral plane is enacted through emotions and desire, fantasy and sentience. It is the plane upon which the vast majority of humanity has its center of gravity, and where it spends significant amounts of its collective dream life. The UR is enacted through the emotional body, and through the limbic system in the brain. The LL comprises the collective culture of all interrelated emotional processes that humanity and the animals of the Earth share in what Wilber describes this as the typhonic, archaic and magic structures. And the LR is composed of the collective matrix of all of these emotional energies.

The physical plane (infrared): The UL consciousness of the higher, subtler level of the physical plane, which the Trans-Himalayan teachings label the etheric, is enacted through an instinctual, impulse-centered consciousness that provides the bridge in communication between the astral, mental and higher forms of consciousness, and the physical body. The UR of the subtlest layer of the physical plane is enacted through the etheric body, which contains the chakra system through which the energies of the various levels of our being flow. The energy of the etheric level of the physical plane is the densest aspect of what the Chinese and Indian systems describe as chi or prana, respectively. This can be understood as the energetic precursor of blood as it flows through the physical body's circulatory system. The LL of the etheric level of the physical plane is enacted through the collective culture, or shared relational space, of what Wilber calls the uroboric, locomotive, and vegetative forms of consciousness, as they express through humans, animals, and plants. The LR is enacted in the collective matrix of all etheric level energies.

As Wilber has noted, the etheric level is what physicists have discovered as the quantum field. The quantum field is understood to be the vacuum state that gives rise to such particles as quarks, electrons and protons. It has been described as a vast field of potentiality that gives rise to the entire physical kosmos. Indeed, quantum physicists have reported that the amount of energy that resides in this quantum state within a single hydrogen atom, for instance, is more than that extant in all known stars. As a result of these findings, many in both the spiritual and scientific communities have started to suggest that this quantum field should be understood as equivalent to the Absolute Reality out of and as which the entire kosmos arises. This approach presents significant problems when we think it through and try to integrate it into a coherent cosmology.

If we did, for instance, consider the quantum field as equivalent to Source we would be forced to consider the great chain of creation in terms of Absolute Reality first emanating physical forces, particles and properties that configure to produce the basics of physical reality. Then we would have to fathom that from this physical reality that is the first emanation of God, on some particular planets self-conscious beings such as ourselves grow and evolve into more and more complex expressions. These move from physical to emotional to mental to buddhic and so on, but each time moving further away from those basic physical properties that are, according to this model, closest to God. This necessarily entails a perspective that sees evolution driving off into the distance away from the divine rather than closer to it, which is, as Wilber has noted, problematic to say the least.

In company with Wilber, we suggest that the conflation of the quantum field with Absolute Source is an error born of flatland thinking that does not recognize or differentiate the multidimensional layers of the kosmos, and the creation process whereby Source emanates the universe. The very fact that the quantum field has measurable, detectible properties discloses its nature as not identical with radically unqualifiable Source, but rather as *a particular layer* of the manifest, conditioned kosmos. Rather than understanding the quantum field as equivalent to the radical Absolute Reality, a more nuanced, rational and sophisticated interpretation would point to the etheric plane. Just as the science on the quantum field shows, the etheric levels of the physical plane are the energetic matrix out of and from which the forces and properties composing the physical dimension field of the kosmos arises, just as the contents of the etheric plane arise from the astral.

The lower levels of the physical plane are the gross, physical levels that we perceive with our five senses. The UL is enacted through the somatic consciousness of the body, which is below the surface of awareness for the majority of human beings[54]. The UR is comprised of the physical body, and the UL somatic consciousness is centered neurally in the reptilian brain stem. The LL is enacted through what Wilber describes as the protoplasmic and physical pleromatic structures. And the LR is enacted as the shared, apparently solid and physical world in which all human civilization exists.

2D Stack Models or Interpenetrating Spheres?

Something to remember is that the above two-dimensional 'stack model' of the planes (with one plane on top of the other) is used only for ease of explanation, and is problematic if it is taken too literally. Unfortunately two-dimensional stack models, which are great for easy-to-understand tabulations, are extensively pervasive in the world of spiritual education and are quite tricky owing to their implicit message that spiritual is 'up there', whilst the 'non-spiritual' is 'down here'.

A more accurate approach to understanding the planes is to visualize them as holonic *interpenetrating spheres*, where each higher plane not only transcends and includes but actually interpenetrates the previous exterior plane. This reflects Wilber's principles of enfoldment and unfoldment, where each higher holon enfolds its junior holons while allowing higher holons to continue to unfold. If we picture it as such, and in relation to the kosmic physical plane only at this stage, we see that the soul planes do not just transcend and include the personality planes, but actually interpenetrate them. The same can be said of the monadic planes, which not only transcend and include the soul and personality planes, but interpenetrate them too. This goes for the kosmic planes beyond the kosmic physical also, which transcend, include and interpenetrate all the way through physical manifestation.

In this sense we could point out that it is not that the soul and monadic level realities are 'up there', far and away transcendent from where we are 'down here'. Rather, it is just the case that their respective enactments of consciousness, body, culture and collective energy-fields have not yet manifested on the most densely embodied levels of Earth. As we will go on to see in this book, it is this process in which the contents of the higher, more kosmically integrated planes, in company with the continued unfolding of radical awakening throughout all of humanity, is described in detail in the Trans-Himalayan teachings. There, this process is understood to be crucial to the bridging of the

---

54. Certain spiritual paths do train individuals to develop conscious awareness of the somatic level, such as shamanism.

gap between humanity, the Earth and kosmos, and the dawning of an increasingly enlightened planetary civilization and culture. We could view the interpenetrating sphere model as shown below.

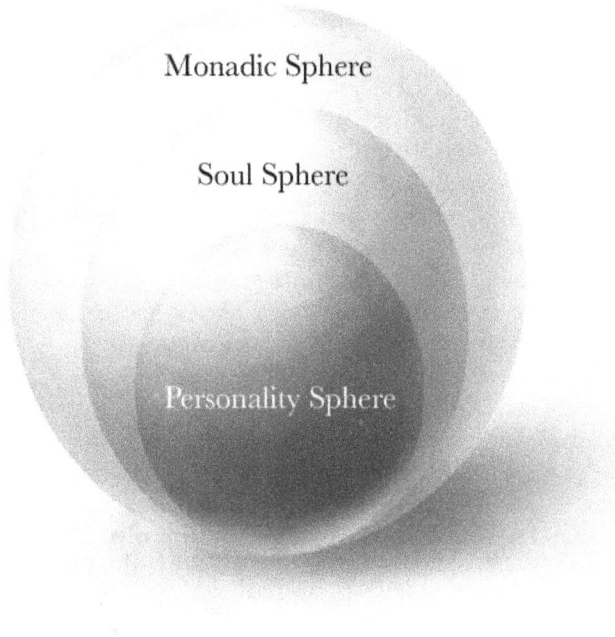

Monadic Sphere

Soul Sphere

Personality Sphere

Figure 16. Graphic showing the personality, soul, and monadic planes in a form where the higher planes transcend, include, and interpenetrate the lower planes.

The Planes are Alive

One of the central transmissions of the Trans-Himalayan teachings is that we live in a sentient, multidimensional universe in which *everything*—not just consciousness but all energy-matter of all levels and grades, from the rocks and stones of Earth to planets, solar systems, galaxies and beyond—*is alive*. The planes are no exception in this regard. Indeed, each plane is understood to actually be the body of manifestation of a great and profoundly evolved deva.

Teachings on what are termed devas in the East, and angels in the Middle Eastern traditions, have a long history. In some respects, accounts of these beings differ, but for the most part there is a strong degree of corroboration across the traditions on how these beings are understood. The word deva has its etymology in the Proto-Indo-European word *deiwos*, which is an adjective meaning "shining", as in "shining beings". Within the Buddhist teachings, devas are considered to be entities on subtle planes of energy-matter who are only perceptible to those who have opened up inner faculties of sight. Such beings may or may not be radically awake, they exist at various levels of evolutionary development and are still subject to birth and decay though according to much longer periods of time than humans. In the Hindu tradition, the devas are considered as supernatural beings of many different classes who dwell on subtle planes as the maintainers and custodians of the environments of those planes, as well as beings who embody and direct the forces of nature. In the Abrahamic traditions of Judaism, Christianity and Islam, there are numerous accounts and teachings about angels, which normally figure as holy messengers and helpers to humanity and the world, clothed in radiant light.

In the Trans-Himalayan teachings, and as was mentioned in the previous chapter, devas are understood to be beings who may, in their most advanced brackets, incarnate through entire planes or express on different planes. The devas are understood to exist in the following classes: pre-conscious though sentient, self-conscious, and super-conscious. All classes are understood to develop and evolve along a substantially different line of evolution to humanity. The Trans-Himalayan teachings suggest that human beings develop along the masculine line of agency, learning to shape energy-matter on the different planes according to purpose, will and consciousness. The devas, as the life and sentiency of the UR bodies and LR energy fields, develop along the feminine line by deepening their capacity for generative nourishing, embracing, and flowing with consciousness and will. One might thus imagine that the bodies of the devas incarnating through the higher planes are more developed along their respective line of evolution. This is owing to them demnstrating greater fluidity in their flow with the Will and Purpose of monadic Life, and the spontaneously arising currents of Absolute Reality itself. It is also possible that the devas incarnating through the denser planes *may* be more developed than those incarnating through the higher planes. Such beings may have made the sacrificial choice to incarnate through the lower planes because the capacity required to enter those fields for the liberation of the beings they contain is greater.

Chapter 10
Identification
Jon

The infinite Conscious Light that is Awakened Awareness and your true nature exists as an unqualifiable, perfect totality, beyond the limitations of linear time and dimensional space. Radical awakening reveals this Boundless Immutable Principle of Reality, in which the coming and going of events in time—the falling of a leaf to the ground, the birth and death of loved ones, stars moving into supernova and the arising and dissolution of universes—is all simultaneously emanating as radiantly Nondual Presence. As Awakened Awareness emanates as the Causal state, Subtle state, and Gross state vantage points, however, time and space arise, and it becomes proper, relatively, to explore the creation process of *how we came to be*.

Play is so much more fun when we have others to play with; in the Trans-Himalayan teachings, it is the will toward Self-perception that arises within Absolute Source that allows it to reflect its own Original Face infinitely and thus to blissfully know itself through the universe. It is this will toward Self-perception as it unfolds through kosmic creation that allows a process of what the Trans-Himalayan scholar and teacher Michael Robbins calls "emanative-Self-division".[55] By this process, Awakened Awareness steps its expression down through the Causal to Subtle to Gross state/phases of creation, wilfully dividing Itself through emanation so as to reflect Itself to Itself.

This process allows the infinite radiance of Ultimate Reality to emanate first as what the Trans-Himalayan teachings describe as the One Universal Monad. This is the Great Breath described in Helena Blavatsky's Secret Doctrine; the electric pulse of the beating Heart of Reality—the exhalation of God so as to breathe out the universe. It is

---

55. Robbins, 2004.

this One Universal Monad that that is Reality's first Self-perception, which then rays forth the galactic monads that incarnate through entire galaxies. According to the Trans-Himalayan cosmology, these then emanate super-constellational monads, which incarnate through systems of constellations. These in turn emanate constellational, solar, planetary, and intra-planetary monads, which all incarnate through their respective fields, until and beyond that 'point' at which we have what we understand as a 'human monadic essence'. This, when expressing on the seven planes which form the kosmic physical plane, has the opportunity to do so through three primary altitudes of identification—the personality, the soul, and the monadic bodyminds.

We can vision this in the symbol of the Tree of Life in the Garden of Eden—a divine tree whose roots and trunk grow out of the rich, black soil of the Ground of Being as an overflowing of its pure fertility. That trunk—the One Universal Monad or Life—then begins to separate into scaffold branches, from which lateral branches grow outwards, from which stems extend and upon which grow untold numbers of leaves. There is no part of this tree that is more 'tree' than another (all monads of all scales are of the same essence - the life-force of Ultimate Reality), and it does not make sense to ask if there is any part of it more connected to the Ground of Being from which it has sprouted. And yet still, the leaves would have a journey of many stages to trace their source back to the trunk.

Correspondingly, the capacity of a human monadic essence, once it has mastered the entire kosmic physical plane, to trace back its line of emanative identification into the One Universal Monad serves as the basis for understanding the kosmic paths of return that we go on to describe more fully later.[56] For now, the focus of our exploration relates to these three major structures or altitudes of a monadic essence's identification with, and expression through, a particular bodymind on the kosmic physical plane.

---

56. This should not be taken to mean that a monadic essence needs to trace it's line of kosmic emanation all the back to the One Universal Monad before it can then radically awaken to Source—Awakened Awareness. Since the latter is the essence and basis of all scales of kosmic incarnation (just as every body of water, from a stream to a river to the ocean is wet) radical awakening can unfold at any stage of evolutionary development where there is enough self-reflexive consciousness to recognize it. This is a topic we go into more detail with later.

The Altitudes of Identification in Detail

For clarity, each of the three levels of identification I will speak about on the kosmic physical plane can be understood as arising from the confluence of a particular vantage point of awareness and certain evolutionary planes. For instance, the personality level of identification arises from the confluence of the Gross vantage point and the personality planes. The soul level of identification arises from the location of a Subtle vantage point on the soul planes. The monad arises from the expression of a Causal and Nondual vantage point of awareness on the monadic planes. Each of these points of confluence between vantage points of evolutionary planes is described as a level or altitude of 'identification'.[57]

Different domains and disciplines on the planet have generated wisdom on each of these levels of identification—personality, soul, and monad—over time. Wisdom concerning the subtler altitudes of identification available on the kosmic physical plane (the soul and monadic bodyminds) has primarily been a contribution of the spiritual lineages of the Earth. Knowledge concerning the more embodied and embedded aspects of identification, such as the personality bodymind, has come from the West. This includes such disciplines as anatomy and physiology, Western medicine, psychoanalysis, behaviorism, cognitive and affective neuroscience, developmental psychology, and social psychology, to name just a few.

When correlating the four sub-lines of the evolutionary awakening vector with the quadrants of an AQAL matrix, we see identification correlating with the UL and UR quadrants (individual interior and individual exterior respectively). We see relationship, and the various planetary cultures of lives operating on different planes through which relationship occurs, correlating with the LL quadrant (collective interior). We see the energetic auras and fields of those communities of lives, as well as access to the planes on which they are expressing, correlating with the LR quadrant (collective exterior). In this chapter,

---

57. In the Trans-Himalayan teachings, the word 'Identification' normally has a very specific meaning relating to either radical awakening, or evolutionary awakening into the monadic bodymind and beyond. In the teachings up until the third phase collaboration between Bruce Lyon and Djwhal Khul, these two are not clearly differentiated. The reason for this is that radical awakening to Absolute Reality, is necessarily entailed in evolutionary awakening into the monad. We feel that the use of the term only for shifts in bodymind identification along the vector of evolutionary awakening is justified owing to the top-down perspective we hold where every depth of identification (monadic, soul, personality, physical bodymind) is a frequency of emanation of Absolute Source.

we will focus primarily on the UL and UR aspects of identification at each altitude available on the kosmic physical plane—the personality, soul and monadic bodyminds. We will leave the LL quadrant to be discussed in Chapter 11, and access to the LR energy fields to be discussed in Chapter 12. For the purposes of clear explanation of the esoteric anatomy of these different layers of bodymind, we will explore the depths of identification from those with which we are more obviously familiar to those that are more esoteric. This order of presentation also represents that according to which the opportunity exists for shifts in identification back into deeper and deeper altitudes of abiding; namely, from the personality to soul to monadic bodyminds, and beyond.

In this exploration we should also note that phenomenologically these shifts often follow a 3-2-1 pattern; that is, a higher altitude of identification is often viewed initially as a 3rd person 'it'. Then, as one comes into relationship with it, it becomes a 2nd person 'other', and finally it is stabilized into as a 1st person altitude of identification.

Before closing this section, it is useful to link this line of evolutionary awakening with Integral Theory and conventional psychological research. In this light, *identification* as I describe it here seeks to access the same content as conventional research that focuses on self-identity (sometimes called 'ego-development'). For example, in the altitudes of development specific to the personality level of identification, the work of Jane Loevinger and Susanne Cook-Greuter have articulated the territory excellently. The perspectives I introduce here extend this work beyond the cognitive focus on the mental planes to show the further reaches of this line. In this way, what I offer here should be seen as complementary to, inclusive of and extending self-identity research from the personality altitudes into the soul, monadic, and kosmic.

The Personality: Physical-Etheric, Emotional, Mental

The first and most embedded layers of identification we will explore are those with which we are most familiar—the physical, etheric, emotional, and mental aspects. These together, when healthy and balanced, form what we call an integrated personality. It is significant to recognize that this actually involves a relatively high structure-stage of development, one to which the majority of humanity has not yet attained. Indeed, it is one that may not fully come online for some spiritual practitioners until some of the deeper layers of identification on the kosmic physical plane—soul and monadic—have begun to express, as well as some degree of opening to radical awakening.

Here, on these most surface altitudes of our being, the Gross state of the Absolute is most pervasive in terms of the way we understand ourselves and the universe as a whole. To begin our explorations here, we will start with the UL interior of the physical, then etheric, emotional and mental aspects, and will then look at their UR correspondences.

Individual

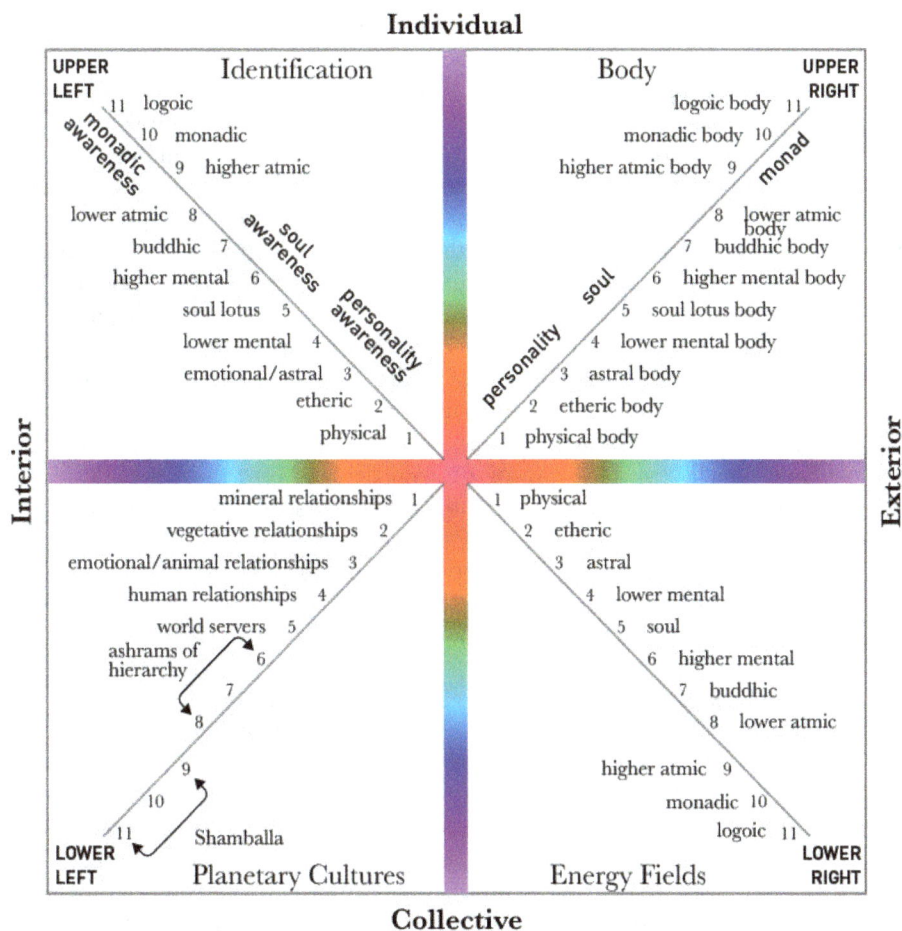

Figure 17: A multi-dimensional consideration of the Four Quadrants of a Human Being.

The UL Interiors of the Personality

Each of these levels—the physical, etheric, astral-emotional and mental—has a unique interiority and sentience. At the physical, non-conceptual altitude of our being, this expresses itself in the physical body's ability to carry memories, traumas and joys in the muscles, joints, bones and cells of the physical body. It expresses in the physical body's innate gravitation toward closeness and physical communion with other physical bodies. It resides in the body's instincts in physical movement such as dance and various forms of exercise; in the natural intelligence for healing; in its natural desire for and sense of deep communion with the Earth, whether that is the sensation of warm seawater upon the skin, sand underfoot, sitting within the tall embrace of trees, or running in open plains. Other examples of expression of this somatic sentience are in its instinctual understanding of and relation with sexuality, and in what has been called the 'wisdom of the body', or the innate capacity for non-conceptual intuitive insight (unencumbered by serial processes of rationality) that can arise from the sentience of the body.[58]

At the etheric altitude, we take our first jump from the solidity of the physical into an ocean of pranic energies, Earth energies, and sexual (libido) energy. The UL interiority of etheric sentience here brings instinctual awareness of the energetic 'tone' or type of vibration being held and transmitted by people, objects, or places, for example. This is the altitude at which our interior sentiency relative to our etheric chakric system is found. We will speak of the chakras more fully later on.

The sentience of the emotional depth of our being has been the subject of extensive inquiry and research, particular by such psychological schools as psychoanalysis, behaviorism and humanistic psychology. From psychoanalysis, we have come to understand our emotional UL interiority as a largely unconscious matrix of emotional psychological material, non-rational, non-verbal, though expressing itself in image, symbol, visceral feeling, fantasy, dreams and associations. Psychoanalysis has brought great insights concerning this altitude of our being and its unconscious, sexual, aggressive, care, nurturance, and/or pleasure-related drives. From behaviorism we have come to understand the emotional sentience as fundamentally expressed through an instinctual *approach- or withdrawal-response* to all stimuli in its environment. The Buddha gave form to this truth some

---

58. An amazing example of this is the documented ability of certain individuals who have cultivated the capacity for deep identification with the consciousness of the body to locate areas of water in drylands.

2,500 years earlier in his teaching on the Four Noble Truths.[59] From the psychoanalytic school we have come to understand how liberating and healing it can be to bring conscious awareness to our often unconscious emotional processes. From the behaviorist school we have grasped that our learnt habits of responding to situations are not set in stone. Rather, we can bring love to our instinctual patterns of response so as to liberate them, gradually and through conscious, sensitive and deliberate exposure to that which they fear or hold reactivity around. As we do this, more and more of the fear and tension we may be carrying can be released, so that each and every current of energy and electrochemical signal in our astral, etheric and physical bodies can increasingly trust.

The emotional nature is a major seat of what Jung called the shadow. As Dustin described earlier, the shadow is that part of ourselves whose self-protective and reactive motivations and drives operate without consciousness to the detriment of our dignity, integrity and overall health. Shadow expressions often take the form of cynicism, pessimism, fear, anger, or jealousy, perhaps, but all are surface expressions of deep wounding.

In each of our trans-incarnational stories, each of us will often carry wounds, scars and pain that only the unconditional love of deeper layers of our nature and all Reality can heal. Truly, we are Christ to our own humanity, and it is through love's capacity not to impose itself or any specific vision on the wounds of our personality, but simply to hold space and listen to anything the emotional consciousness needs to express or release into the light, that the needed healing can occur. It is a process of unconditional *befriending*, and it is one that is deeply important. Through it, the light of awareness is brought to warm and illuminate all those areas of our emotional consciousness that have remained for so long in darkness. The personality as a whole begins to be able to support the full incarnation of the deeper aspects of our nature, and the doors of Earth are opened to love.

Two schools that lend a great amount of light to our understanding of the UL interiority of the mind are cognitive psychology and such schools of philosophy as rationalism and logic. From these we have come to understand mental consciousness as fundamentally related to how human beings process, encode, retrieve and work on information. In fact, within the discipline of cognitive psychology, the most prevalent analogy for the workings of the cognitive mind is a computer, processing information using both parallel and serial operations to take it in,

---

59. The Four Noble Truths taught by the Buddha were 1) the truth of suffering (the word used is *dukkha,* which can be translated as 'incompleteness', 'anxiety', 'dissatisfaction', or 'suffering'); 2) the origin of suffering; 3) the possibility of the end of suffering; 4) the Noble Eightfold Path that leads to the end of suffering.

recognize its significance and act upon it. The analogy here is that our mental processes (UL) resemble software run on the hardware of the brain (UR). Indeed, it was on the pioneering work of Alan Turing, among others, who laid the first solid ground of our understanding of information technology, that cognitive psychology was reared.

Over the course of growth, mental consciousness learns to wield a blade of differentiation, clarity and discrimination, whilst integrating a capacity for inclusion and expressed care. It is on this path that we suggest that mental consciousness moves through the structure-stages described in Integral Theory and Spiral Dynamics. This extends from red to turquoise, which structures are inclusive of the personality level lines of development discussed by such researchers and theorists as Piaget, Loevinger, Cook-Greuter, Vygotsky and others. Dustin goes on to describe these stages of intelligence unfoldment they are understood in Integral Theory in Chapter 14.

Fundamentally, as mental consciousness develops, it allows ever more sophisticated expressions of perspective taking. This begins in an egocentric phase of growth, where only one's own first person perspective is seen be relevant. As development proceedes, if the supporting conditions are favorable, an individual moves on to develop the capacities for second person empathy and third person objectivity. Beyond this, and as we move into higher structure-stages of intelligence unfoldment, we become able to take 4th person, 5th person, 6th person perspectives and beyond,[60] each taking more and more beings and components of whatever systems our cognition is working on into consideration.

Collectively, and as previously stated, these four aspects—physical, etheric, emotional, and mental consciousness—form what we call the consciousness of *the personality*. Over the course of growth and development, the personality becomes increasingly integrated and each of these aspects becomes a riverbed for the deeper layers of identification and Reality itself to pour their illumined mind, heart, wisdom and enlightened will into. As this process deepens, it allows personality consciousness, which at its height is developed to the level of worldcentric *concepts*, to start becoming an actually sentient, conscious,

---

60. Whereas a third person perspective begins to objectify simple content in the form of concepts, a fourth person perspective can objectify the concepts themselves. Once concepts are objectified, a fifth person perspective can string concepts together and begin to see how they relate to each other in a more systemic fashion. This level of perspective taking can hold two seemingly opposite concepts at the same time in paradox, without contradiction. Beyond a fifth person perspective, a sixth person perspective taking capacity relates to the capacity to compare systems to other systems. The sequence of increasing complexity continues indefinitely.

worldcentric *experience.*[61] In this, all of nature—the noosphere, biosphere, and geosphere begins to reveal itself as Nature—the living, breathing body of the Divine.

The UR Exteriors of the Personality

In terms of the UR quadrant of these depths of personal identification, we speak here of the physical, etheric, astral (emotional) and mental bodies.

In relation to the physical body, modern medicine, anatomy, physiology, neuroscience, cardiology and various other specific disciplines have provided extraordinary insights. The development of these disciplines has been fundamental to our understanding of the spectacular intricacy of function and structure of the physical body and how to maintain its optimal health.

In relation to the etheric, astral and mental bodies, Eastern esoteric traditions such as Vajrayana and Vedanta have contributed considerable wisdom through the expertise cultivated by communities of practitioners, teachers and masters, of psychic abilities or 'siddhis' such as clairvoyance, clairaudience, and psychometry, for instance.

Each of these bodies, which compose the exterior of the personality, is understood to exist in various frequencies of energy-matter. The physical body exists in solid (e.g. bone, organs, and muscle), liquid (the various bodily fluids) and gaseous (oxygen, carbon dioxide, nitrogen, and some neurotransmitters) matter. The etheric body exists in etheric energy-matter, which is understood to be conventionally non-perceptible to the senses. The astral-emotional body exists in an even finer grade of subtle energy-matter. The mental body again exists a lighter and faster rate of subtle energy-matter than the astral.

According to the Tibetan and Indian traditions, within each of these subtle bodies (etheric, astral, and mental) reside chakras. These chakras exist for the reception, circulation and potential transmutation of personality, soul, monadic, kosmic and Source energies throughout our system. The various structures of the circulatory, respiratory and central/peripheral nervous systems in the physical body (such as bodily organs, veins, and nerve fibers, etc.) allow circulation and communication of needed energy, electrochemicals and nutrients; in the same way, the anatomy and physiology of the mental, astral and etheric bodies resonates with the same living pattern. The etheric, astral and mental bodies are understood to exist as a living fabric composed of millions of

---

61. Wilber, 1995.

energetic threads along which force, light and information flow. Within those bodies there exists a major and minor system of chakras where vast numbers of nadis—or energetic threads along which information and energy meet around a central laya—center, or 'zero-point' at the heart of the chakra. In the Trans-Himalayan teachings, the major chakras are seven, and are found at the crown, brow, throat, heart, solar plexus, sacral, and base of the spine.

Apart from these there are many other chakras throughout the bodies, such as at the palms of the hands, the soles of the feet, the genitals, the shoulders and areas of the face, among others, which constitute just a few of the focal points within the minor chakra system. Furthermore, each of these chakras is understood in the Trans-Himalayan teachings to have layers or tiers to its form corresponding to the personality, soul, and monadic altitudes of identification. As is described by the Eastern traditions, there is a profound nature-teaching to be found in how the structure of each chakra, in terms of the energetic currents and channels of flow composing them, takes the apparent form of a lotus with three layers of petals and a central core. This structure of each of these chakras represents a microcosm of the greater soul body, which we shall go on to describe later.

The outer layer of petals corresponds to that level of a chakra attuned to personality unfoldment and the Gross state, and it is during the course of personality development and integration that this layer of the chakra lotus is activated and subsequently unfolded. The middle tier is attuned to soul altitude unfoldment and the Subtle state, and thus is activated and unfolds more and more as the soul altitude of our being is actualized. The inner tier is attuned to the monadic depth of identification and the Causal state, and becomes activated and unfolded as that depth of identification comes into expression. At the zero-point core of each chakra there is the vantage point of empty Awareness, and as radical awakening opens up and the Nondual Awakened Awareness that is the basis of all these layers is recognised, the chakras bloom and become transmitting stations for Reality's Life-force. Naturally, as this occurs, the energy of the Absolute that has been (from the relative perspective) trapped in the energetic and dense physical bodies through self-contraction and non-recognition of inherent Freedom is able to shift, move and be released. This relates to the kundalini processes that come with radical awakening, and that facilitate the emergence of a progressively transparent bodymind through which Absolute Reality can expresses freely.

Each of the bodies, etheric, astral and mental, has its own system of chakras, with the nadis that compose it both residing as the *fabric* of that body, as well as *interconnecting* with the nadis composing the other bodies. It is through the millions of connections both within and between the nadis of the etheric, astral and mental bodies that we find the objective (UR) understanding of how our various forms of energetic, emotional and mental behaviors are supported, learned and engaged. This closely reflects how learning is understood to occur within the physical body, and particularly the brain. Cognitive neuropsychology, most especially in the clinical field, has documented a strong degree of anatomical specificity in the brain in terms of the degree to which seemingly discrete neural circuits and areas of neural anatomy often support very particular behaviors. Patients with lesions in certain brain regions through accident or disease often demonstrate deficits in specific behavioral skills, and yet maintain perfect functioning in other areas. One such case, that of a patient known in the literature as L.M., involved a syndrome known as akinetopsia,[62] where the patient was unable to perceive movement or objects in motion. Fascinatingly though, all other visual faculties (color perception, spatial perception, orientation, retinal disparity etc.) remained unaffected. This implicates the anatomical separation between different visual faculties as responsible, with damage having occurred to one of them that is specifically responsible for motion perception and nothing else. If one considers briefly just how many types of behavior are as little noticed but as functionally important as that which was damaged in this case, one may begin to grasp the number of functionally distinct brain regions and neural connections within the cerebrum (which some have calculated as 10 to the 83rd power!).[63]

Over the course of development, the brain is able to act increasingly as a synthetic instrument in which all aspects of a human being's constitution—personality, soul and monad—may be supported for expression on the physical plane. It therefore seems reasonable that the structural and functional organization of the brain reflects, as the *surface* structure and in one coherent anatomical locus, the *deep* structural and functional pattern of organization demonstrated by the subtle bodies also. If we break this down, we can first point out that within the brain particular behaviors are supported by different brain regions and neural connections. On this basis, it would also seem reasonable to suggest that within the subtle bodies, particular behaviors that are eventually biologically, neuropsychologically and behaviorally rooted in the brain have their source in specific centers and pathways of etheric, astral and mental energy. We could thus suggest that the structural and functional complexity of the centers and pathways in the subtle bodies must be at least as complex as it is in the brain. Indeed, this hypothesis is supported

---

62. Zihl, Cramon, & Mai, 1991.

63. Norfleet, n.d.

by the description of the etheric body in Indian Hindu teachings as composed of "millions of tiny streams or lines of energy."[64]

In a manner that provides an esoteric perspective on Social Learning theories and their neurological extensions, some of these insights on how development happens through the energetic and physical bodies can be extended further. Just as modern day cognitive neuroscience has been able to reject strong arguments of determinism through the recognition of the brain's startling degree of neuroplasticity, or its capacity to reorganize itself and change through experience, so also is it likely that these etheric, astral and mental energetic pathways that support particular behaviors are also highly plastic. That is, they are capable of strengthening, diminishing, re-wiring and radical reorganization, and this is the physical basis of the awakening process along both the radical and evolutionary vectors.

The relative size of the bodies is a final point worth touching on. The Eastern teaching understands each body subtler than the physical to actually transcend, include and interpenetrate each denser body spatially. That is, each subtler body extends like a halo of variously colored light and energy-matter around the periphery of the more dense body, containing that body within it and energetically interpenetrating it—very much like Russian dolls. Therefore, the etheric body is understood to extend in its radiance beyond the physical body, include it in its light, and interpenetrate it as a web of energy-light-matter superimposed upon the dense physical structure. So also with the astral body again, which is understood to transcend (extend beyond spatially), include and interpenetrate the etheric and physical bodies. Then again this will occur with the mental, which spatially transcends, includes and interpenetrates the astral, etheric and physical bodies. This process will then be repeated with the soul and monadic bodies too.

One thing that happens in advanced stages of development, in relation to the personality, soul and monadic altitudes, is that as a being develops intelligent inclusivity, love, wisdom, and identification with divine Will, Purpose and Reality-transmission more and more fully, their energetic presence begins to extend wider and wider. As a result of this process, they become able to include, nurture, love and empower more and more beings within their fields of contact.

This speaks to how energy, emotion and mind are not left behind as spiritual development increasingly unfolds, but rather the etheric, astral-emotional, and mental bodies of masters grow so as to extend beyond, include in nurturance and empower through interpenetration more and more beings. Their etheric bodies hold more and more beings in a sphere of their energetic presence. Their emotional bodies leave the

---

64. Bailey, 1950.

state of emotionality and enter the world of increasingly unbounded sentience. And their mental bodies are able to work increasingly with concepts and structures of thought which are universal in their inclusivity. From an UR perspective, these mental structures reside on the mental plane as energetic bodies whose architecture and geometry is so all-encompassing that whole domains of global thought are able to be integrated within them.

This leads eventually to the phenomenon we shall explore extensively in *Earth is Eden*. According to the Trans-Himalayan teachings, eventually beings have the opportunity to open and evolve into stages of mastery where their energetic presence, love, and will-to-freedom for all beings are expanded to profound scales. Such being then extend their spheres of love, care, and responsibility so widely that they are able to incarnate through and ensoul entire soul groups, planets, solar systems or galaxies as their bodies of manifestation, and all of the evolutions living within them.

The Soul

According to the Trans-Himalayan teachings, at the second major depth of our nature we are souls—beings of innate love and wisdom— journeying over aeons through the Earth School (and other planetary schools as we will go on to explore later) into greater and greater light. Just as the personality is understood as a threefold being in the Trans-Himalayan teachings, so also is the soul. Specifically, the soul is understood to be composed of the higher mental, buddhic, and lower atmic principles, which translate as illumined mind, intuitive love-wisdom consciousness, and enlightened will respectively. In the second and newly emerging third phase of the teachings, a cosmological analogy is drawn between the consciousness of the soul and the golden light of a sun. As souls we learn to radiate love and light without question and hesitation to all beings, just as the sun unconditionally radiates light, heat and magnetic stability throughout the solar system.

It is important to note that by the soul we do not understand just a more spiritual personality, but a literal *light-being* whose consciousness and energy are permeable and transparent to all other souls. Souls are characterized as interpenetrating spheres of consciousness (UL) and energy (UR), each holographically identified with the nature and trans-incarnational journey of each. As the Trans-Himalayan teachings describe within soul consciousness, the reality of our interconnectedness, interdependence, and collective story of becoming is so clear that unconditional love and wisdom are seen as "pure reason".[65] Here, the Subtle state of the Absolute is most pervasive, making the soul planes shimmer with vast light, sound, and rays.

The UL Interior of the Soul

As Wilber has noted, this is the depth of pure love-consciousness at which truly selfless and compassionate morality is resurrected as intrinsic to any relational exchange.[66] This isn't a 'soft' morality though. The eye of the soul sees from such a depth of surety in the love, light and goodness of Reality, as well as the immortal nature of our being and the great tidal cycles of evolution that all lives are passing through, that its expression of love may be deeply challenging at times. To the consciousness of the soul it is not the pain, joy, hurt, fear, and successes of life which matter. Rather, it is the light that can be birthed into our world through them, and the love and honoring of all of life that can be grounded by meeting them openly, honestly, receptively, and in fullest dignity.

Often, the soul is described as the 'transpersonal self'. The use of the term transpersonal is not because it is categorically 'not-personal'. Rather, just as the deeper and more inclusive (and yet still relative) reality of the monad *transcends and includes* the more relative reality of the soul, so does the latter transcend but include the personality. It transcends it insofar as it is *more* than the personal ego, *deeper* than the personal ego, *bigger* and *fuller* than the personal ego. As Djwhal Khul has said, "…this one life is but a second of time in the larger and wider existence of the soul".[67] And yet its inclusivity is found in the fact that the movements, journey, growth and unfolding of the personality bodymind are all part of the soul's multi-incarnational journey and are thus, simultaneously, utterly sacred in their every contour to it.

While the awareness of an integrated personality eventually embraces a planetary (worldcentric) level of scale, the consciousness of the soul eventually spans a solar level of scale. Indeed, as both Djwhal Khul's teaching and the transpersonal hypnotherapist Michael Newton's research suggest, not all souls on Earth at present originated or have always been here. In his work with Alice Bailey, Djwhal Khul described how the current collective of souls on Earth is composed not

---

65. Bailey, 1960, p. 27-28.

66. Wilber, 1977.

67. Bailey, 1957, p. 7.

just of souls for whom their 'home world' is the Earth, but of a smaller number of souls from other planetary worlds too.[68] Here is Michael Newton:

> In the early days of my studies of souls, I half-expected that those subjects who could recall other worlds would say they had lived in our galaxy within the neighborhood of the sun. This assumption was naive. Earth is in a sparse section of the Milky Way with only eight stars that are ten light years from the sun. We know our own galaxy has more than two-hundred billion stars within a universe currently speculated at one-hundred-billion galaxies. The worlds around the suns which might support life are staggering to the imagination. Consider, if only a small fraction of one percent of the stars in our galaxy had planets with intelligent life useful to souls, the number would still be in the millions.
>
> From what I can gather from subjects willing and able to discuss former assignments, souls are sent to any world with suitable intelligent life forms. Out of all the stars which are known to us, only four percent are like our sun. Apparently this means nothing to souls. Their planetary incarnations are not linked to Earth-type worlds or with intelligent bipeds who walk on land. Souls who have been to other worlds tell me they have a fondness for certain ones and return to them (like Earth) periodically for a succession of lives.[69]

The UL consciousness of both Earth souls and souls from other worlds eventually embraces a solar scale of awatreness, communion, and activity among soul beings throughout the solar system. And as a being's continuity of consciousness is established between their regular waking personality consciousness and that of the soul, so also unfolds the potential capacity for conscious recognition and engagement with these soul altitude experiences of intra-solar system communion, communication and extra-planetary travel on soul planes. A further quotation from Michael Newton:

> It is important for the reader to understand that when people do recall living on other worlds they seem not to be limited by the dimensional constraints of our universe. When souls travel to planets intergalactically

---

68. Bailey, 1942, p. 201.

69. Newton, 1994, p. 192.

or interdimensionally, they measure the trip by the time it takes them to reach their destinations through the tunnel effect from the spirit world [the soul planes]. The size of the spatial region involved and the relative position of worlds to each other are also considerations. After listening to references about multiple dimensional realities from some of my subjects, I am left with the impression they believe there is a confluence of all these dimensional streams into one great river of the spirit world [the soul planes]. If I could stand back and take apart all these alternate realities seated in the minds of my cases, it would be like peeling an artichoke of all its layers down to one heart at the core.[70]

The UR Exterior of the Soul

In terms of the UR quadrant of the soul, as Wilber noted in his *Spectrum of Consciousness*, this is an altitude of identification, found and experienced as we venture and abstract our awareness ever deeper into the subtlest reaches of the mind and beyond, on whose involutionary arc the soul body- and energy-based boundaries between self and other are not yet fully crystallized. At the altitude of the soul the walls of separation between beings are markedly more fuidic. Here, the fabric of kosmic space upon which the hieroglyphs of Truth are written is more nakedly exposed to the soul's already illumined and intuitive mind, and the awakening soul need only glance into its own heart to know the infinity of Godhead.

In recent decades, a wealth of transpersonal research—psychological and hypnotherapeutic—has emerged to confirm and extend the teaching of the traditions on the reality of the transpersonal soul. In terms of a definition of the soul and its form, garnered from decades of research, Michael Newton writes:

> I see the soul as intelligent light energy. This energy appears to function as vibrational waves similar to electromagnetic force but without the limitations of charged particles of matter. Soul energy does not appear to be uniform. Like a fingerprint, each soul has a unique

---

70. Newton, 1994, p. 197.

identity in its formation, composition and vibrational distribution.[71]

Newton's research, conducted over decades with hundreds of individuals who had no prior contact with each other, has contributed a startling degree of corroboration between individuals on the UL and UR nature of soul consciousness and body, the LL planetary cultures of relationship that it experiences throughout its incarnational journey, and the LR environment of the soul-planes. A central feature of this corroborative testimony is the holarchical organization of souls, in their various groupings within the soul planes, with the altitudes of that holarchical organization differentiated on the basis of soul development. This development is defined according to the depth, or the degree of loving inclusivity and wise discernment of a soul's consciousness, across planes and beings. We could therefore say that the capacity to embrace greater and greater numbers of beings in its consciousness and energetic presence, whilst retaining the capacity for clear sight as to where on the evolutionary chain of hierarchy they stand and how they might be best served, is the central quality for development for the soul. This means that the more advanced a soul is, the more souls and relationships they are able to transcend, include, and stimulate with their sphere of love, wisdom, and energetic presence.

This brings an interesting insight to the perspective held by St. Paul in his conflict with St. Peter over the degree to which older Judaic traditions, rituals and practices would need to be retained in Christianity. Paul, in recognizing the scale of unconditional inclusivity embodied by the great soul of Christ, suggested that subsequent to the crucifixion, where Christ had opened his love for all beings to a global scale, no such practices were necessary. For Paul, the world had shifted into a state of grace in Christ and could no longer solely be defined as operating solely under the Law of Moses. According to Paul, Christ had established such a profound scale of loving presence in consciousness and love-energy throughout the physical, etheric, astral, mental, and soul planes of the Earth that the journey of all younger souls had been forever changed. No longer were the old forms and rituals necessary. From the moment the soul body of Christ had exploded with love for all beings to an unprecedented scale, *love was here*. Now, every particle, atom and molecule of the atmosphere surrounding every individual, animal, plant and stone was vibrating with love-energy to a level never before known on Earth as part of a global field of Christ-presence, and disciples and practitioners of the path could enter into Love *directly*. Esoterically speaking, we could say that from that point onwards, the

---

71. Newton, 2001, p. 85.

body of every soul from then until now and into the future was and is composed in part of the subtle particles, atoms, and molecules of love that formed Christ's own soul body. A profound kosmic nature-teaching for this is found in the increasingly supported theory in astrophysics that the sun and planets of our solar system formed some 4.5 billion years ago from the supernova explosion of another giant, ancient sun in our galaxy, the shockwave from which carried the dust and gas and elements that now compose them.[72]

In the Trans-Himalayan teachings, and as was touched on earlier in the passage on chakras, it is agreed that the intersecting streams of energy that compose the soul body are understood to take the form of a lotus. This soul lotus, in terms of its anatomical connection with the form of the personality, is anchored in the crown chakra. From that energetic altitude, it overshadows the personality for much of its trans-incarnational history. When the continuity in awareness between the soul and personality deepen sufficiently for the soul consciousness (UL) to begin to increasingly express through the personality, however, there are corresponding changes in the soul body (UR) (as well as in its relation to LL planetary cultures and LR fields of interaction). Specifically, the soul body then begins to extend spatially from above the crown of the head progressively further down to embrace the bodies of the personality—mental, emotional, etheric and physical. Eventually it transcends, includes and permeates the entirety of all these, saturating them with its radiant light, unconditional love, and wisdom.

Like the form of each of the chakras as described earlier, the soul lotus-body is described in the Trans-Himalayan teachings as having *three tiers of three petals* which open over the course of the soul's entire trans-incarnational journey so as to reveal a jewel of fiery Life at its core. As we saw at the beginning of this chapter, the monad is understood as an emanation of greater kosmic monads, which are ultimately emanations of Reality. So also are the soul and personality emanations of the monad on the kosmic physical plane. This jewel at the core of the soul is therefore understood as the anchorage point for the stream of monadic Life-force that supports it. The anchorage point for the emanated stream of monadic Life in the personality is in the base chakra as kundalini. This is something we will return to later.

---

72. Did a Supernova Shockwave Create Our Solar System? New Finding Says "Yes", 2012.

Figure 18: An artistic dipiction of the soul lotus by Lynda Vugler.[73]

In the Trans-Himalayan teachings the jewel at the core of the soul is described as a diamond of blue-white energetic fire. Diamonds are forged in the heat and pressure of the depths of the Earth, and it is their unique lattice arrangement of carbon atoms that produces their indestructible hardness, clarity, and strength. Similarly, the diamond of monadic Life at the heart of the soul is unbreakable in its sacrificial will to serve the kosmic Purpose that is being progressively revealed on Earth. Just as diamonds also show the remarkable property of electrical conductivity, it is the role of the diamond at the center of the soul to conduct and transmit the kosmic Will, Purpose and Life-force transmitted from the monad. These are poured forth into the soul and eventually the physical altitudes of our identification, relationship, and environment.

This occurs at a relatively advanced stage of the path both vertically and horizontally. The vertical transmission extends from the monadic

73. Lynda Vugler's work is viewable at: http://www.simplysacredspaces.com/

altitude of identification into and through the personality, particularly to and through the core of each chakra. Horizontally, the transmission extends between souls as part of a living nerve-system on the soul planes. As it does this, the petals of the soul body are increasingly pushed open from the inside out. Eventually, owing to the pressure emanating from the core of the soul and forging ever more powerfully the diamond at its heart, the soul body as a whole is burned clean in the fire of transmission established between monad and personality. This occurs at a stage of evolutionary awakening that we will go onto describe as the 4th initiation, and it leaves of the soul only a diamond of blue-white flame through which the monadic Power and kosmic Light of the universe can pour.

Of the three tiers of petals surrounding the jewel, the outermost tier is described in the Trans-Himalayan teachings as composed of 'knowledge petals', which open as the soul deepens its knowledge and wisdom concerning life, incarnation, relationships and Reality, gathering the fruits of learning from each life like a bee taking pollen from a flower. These are connected to the Gross state and participate most fully in the Gross vantage point. The middle tier of petals are 'love petals'; these open as the soul progressively expands its loving and wise embrace of all beings as was described above. These are connected to the Subtle state, and to the Subtle vantage point of awareness. The inner tier of three petals are known as 'sacrifice petals'. These are connected to the Causal state, and a Causal vantage point. They are the last to open, and they do so as the soul increasingly dissolves its perceived individuality in the river of kosmic Will, Power and Purpose being transmitted from the monadic level. As this process deepens, the soul leans more and more into its own sacrificial heart and will-to-die. It consents to the last remaining and most subtle illusions of separation to fall away so that it may give itself, its knowledge, wisdom and love, to the ALL. As the final sacrifice petals open the illusion of duality is dissolved; all resistances to free expression of monadic Will, Purpose, and Life are increasingly resolved; and the abiding structure-stage of identification moves stably into the monad. This is the release of the soul from its crucifixion upon the Tree of Knowledge of Good and Evil, so that the essence at its core might recognize itself as the Tree of Life spawned from the One Absolute Ground and Base.

In this movement, the entire trans-incarnational journey of the soul that has spanned hundreds of thousands of lifetimes is both synthesized into a single point of uniquely colored light, and given away. No longer is the illusion held that this soul has surrendered into a deeper identification with the monadic bodymind. Rather, now it is experientially recognised that the Boundless Immutable Principle of Awakened Awareness, arising as the monadic Life-essence, has released its identification with the last, most subtle resistance to the Great Perfection. Indeed the last petal of the soul that is said to open is known as "the utter sacrifice of all forever".[74]

Over the course of these lifetimes, it is *through* the incarnation of the soul on the personality planes (mental, astral, etheric and physical) that it engages its path of blossoming and unfoldment, harvesting the fruits of each incarnation in the inter-life state, or what the Tibetans call the Bardo. In the Vedantic teachings, the soul stores the trans-incarnational memory in what is called the *vijnanamaya kosha*, or the 'wisdom-apparent sheath'. This is another teaching on the UR body of the soul, which the traditions understand to act as the reservoir of the fruits of incarnational experience—the karmic seeds of becoming gathered and assimilated in the inter-life state, and then worked out in future lives. That the soul is able to unfold an ever deeper light, love and consciousness through the personality and its degree of responsiveness to the soul points to the profound teaching of the Tibetan Master Djwhal Khul with Alice A. Bailey that "The soul is, therefore, in deep meditation during the whole cycle of physical incarnation", and that "meditation is rhythmic and cyclic in its nature as is all else in the cosmos." On this subject, which has clear significance to spiritual practitioners, Djwhal Khul continues:

> The soul breathes and its form lives thereby. The rhythmic nature of the soul's meditation must not be overlooked in the life of the aspirant. There is an ebb and flow in all nature, and in the tides of the ocean we have a wonderful picturing of an eternal law. As the aspirant adjusts himself to the tides of the soul life he begins to realise that there is ever a flowing in, a vitalising and a stimulating which is followed by a flowing out as sure and as inevitable as the immutable laws of force. This ebb and flow can be seen functioning in the processes of death and incarnation. It can be seen also over the entire process of a man's lives, for some lives can be seen to be apparently static and uneventful, slow and inert from the angle of the soul's experience, whilst others are vibrant, full of experience and of growth.[75]

---

74. Bailey, 1925, p. 824.

75. Bailey, 1934, p. 62.

Supernova

Often in the mystery schools and esoteric traditions of ancient times, such as the Orphic Greek and Egyptian, the soul was symbolized by the sun. Like souls, whose paths of radiant lotus unfoldment comes to its eventual fruition and avalanche into mastery, when suns die they have a number of different potential destinies depending on their mass. If their mass is large enough, they may supernova and collapse in on themselves to leave a black hole. The correspondence to this in the human system is the stabilised transition of identification from the soul bodymind on the soul planes into the monadic bodymind on the monadic planes. If their mass is smaller, however, as our Sun's is, they are likely to end their lives as a white dwarf. We might imagine the correspondence to this on Earth being the choice made by those liberated beings whose soul bodies have been consumed in the lambent flames of their own realization to *not* transition off of the soul planes onto the monadic, but to stay. Such beings retain the anchorage of their identification in the diamond of fiery energy left after the dissolution of the soul lotus, so as to act as conductors and transmitters of the kosmic Will, Purpose and Life of the monad onto and through the soul planes.

As we will go on to explore later in this book, the Trans-Himalayan teachings see the kosmos as a multidimensional living fractal, with the wisdom teachings of every level, from the galactic to the atomic, and from the most subtle to the most gross, mirroring each other. As a kosmic correspondence to the narrative of the soul just described, in 2004 scientists found that when suns expand into the form of a red giant and then pass into the form of a white dwarf, some two billion years later, the carbon in their hearts crystallizes into a gigantic diamond.[76] It is suggested to be likely, therefore, that the future destiny of our Sun is the same. Eventually, all its fuel for nuclear fusion will burn out and the solar system will consume itself in kosmic fire. Then, our Sun will contract into a shining point of white light in the midnight blue of universal space, whose remaining core embers, once gone out, will crystallize into a kosmic diamond.

The Monad

As the first paragraphs of this chapter elucidated, a distinction needs to be made between the monadic essence and the monadic bodymind. At the level of the monadic essence, we are all the emanated Self and

---

76. Diamond Star Thrills Astronomers, 2004.

Life of the One Universal Monad (the trunk of the Tree of Life). The monadic bodymind is the subtlest structure of identification through which the monadic essence expresses on the kosmic physical plane. This monadic bodymind is so subtle that it offers no resistance to the monadic essence's instinctual radical wakefulness as a wave upon the kosmic Ocean of Awakened Awareness—a rush of wind through the vast open sky. At this depth of identification, the Nondual and Causal states/phases of the Absolute shine forth nakedly. Therefore, as monads we reside as innately Self-realised sparks of Absolute Wakefulness and its dynamic energy. In the words of an esoteric text quoted in the Secret Doctrine whose source is so ancient it must remain unknown, and that involves a question asked from a master to a student:

> 'Lift thy head, oh Lanoo; dost thou see one, or countless lights above thee, burning in the dark midnight sky?'

> 'I sense one Flame, oh Gurudeva, I see countless undetached sparks shining in it.'[77]

The UL Interior of the Monad

Earlier, at the beginning of this chapter, we saw how it is the Will toward Self-perception that arises within Absolute Source that allows it to reflect its own Original Face infinitely. This eruption of the Will toward its kosmic Self-reflection could be described as a function of its pure radiance, the first kosmic thrill of which arises as the One Universal Monad. We saw how this One Universal Monad, described in the Indian traditions as Ishvara and in Tibetan Buddhism as Vairochana, then emanates through all octaves of kosmic incarnation, from galactic to super-constellational to constellational to solar to planetary, and then as a human monadic essence.[78]

In the cosmology of the Trans-Himalayan teachings, the UL interior awareness of each monadic essence participates in both the radical Reality of the Absolute Self, and in its Self-recognition as a particular level of kosmic evolutionary emanation. That is, along the radical awakening vector we could say that the interiority of the monadic essence resides in unbroken radical awakening to and as the Absolute

---

77. Blavatsky, 1928, p. 145.

78. The process of emanation of course extends further into smaller and denser forms of incarnation than this, but in this book we only consider the process up until a human monadic essence.

Reality of Unbounded Awakened Awareness. Along the evolutionary awakening vector, in its UL, it self-recognizes as the emanated Self of a kosmic Logos (which is itself the emanated Self of the One Universal Monad at a particular octave of Self-reflection[79]). It abides in continual apprehension of, identification with, and transmission of the energetic current of the divine Will and Purpose of the kosmic Logos within whose aura and body of manifestation it resides.

The UR Exterior of the Monad

In terms of the UR quadrant of the monadic bodymind, owing to it arising as a confluence of a Causal vantage point and the monadic planes, third phase Trans-Himalayan teachings symbolize the body of the monad with the invisible or dark body of a black hole. This is an opening to divine Emptiness with such mass, density and power that it warps all space and time around it in conformity with divine Purpose. At this depth of our being, we are radically awake kosmic Life, embodied in tears of space-time that form the subtlest and most refined sheath to differentiate interior from exterior.

This subtlest depth of identification on the kosmic physical plane transcends, includes and resides as the essential core of the soul and personality aspects of identification in both space and time. Spatially, the monadic body is the dark halo that surrounds, is the womb of and exists as the essential basis of both. Temporally, this is the depth of our being that holds the entire history of our trans-incarnational journey in the dark pupil of its awakened eye. This extends from even before our individualization as a soul that would then incarnate through hundreds of thousands of personalities over time, to the present, and right through and beyond to the height of our kosmic Purpose.

Additionally, like the Hawking radiation emitted from black holes, or the continually radiated informational essence of every particle they consume, the monadic body sounds forth the synthesized note of every personality lifetime and waterfall moment of soul growth as they are consumed over time into its heart. As Hawking radiation from a black hole radiates out mass at a higher rate of evaporation than it is absorbed, eventually each black hole, like every monadic bodymind, will radiate out into the kosmos everything it is and has ever been, thus disappearing and allowing the monadic essence entry into exploration of other kosmic planes. Using the Tree of Life analogy from the beginning of this chapter, we could liken this to when leaves

---

79. We will go on to explore the factor of octaves more fully in Chapters 18 and 19.

fall from the branch stems of a tree to the Ground and are then broken down so that the whole tree can reabsorb their nutrients and life via its roots. Meanwhile, the monadic essence, abstracted from the monadic bodymind, retraces its root deeper into the greater branches.

While the quality of energy particular to the personality is described in terms of intelligence, and that of the soul as love-wisdom, the quality of energy transmitted by the monadic bodymind is divine Power, Will, and Purpose. As such, the Trans-Himalayan teachings symbolize monadic energy in the form of lightning, and its transmission of Truth as thunder.

In Ancient Egyptian mysticism, this aspect of our being was known as the *akh*, which translates as "radiant spirit" or "radiant power". This they understood to be the always-already transfigured depth of our being that was of equal status to the gods and that upon liberation could return to the heavens to take up its place in the body of the sky goddess as a star.

In the Kabbalistic tradition, the monadic essence is described as comprising the *chayah* and *yechidah*, and there are some sublime descriptions given. When defining the chayah we are told, "Chayah essentially means 'living essence' or 'life-force'. At this altitude, one is completely enveloped in the shade of the divine."[80] And on the yeshidah:

> The *yechidah* relates solely to the purpose of creation, and its entire essence is intent on reaching *sham*—the ultimate goal of existence, where all is one and there is only unity. This is reflected in the very name of this aspect of the soul: *'yeshidah'*, meaning 'singular', and 'one'.[81]

Identification with and as the monadic bodymind is described by the Sufis as *'fanah fi Allah'*, or extinction of the self in the Will of God. And in the teachings of the great Indian sage, Sri Aurobindo: "That substance is the self of the man called in European thought the Monad, in Indian philosophy, Jiva or Jivatman...."[82] Aurobindo also describes it the 'central being'. It is the spark of divine essence at the core of every being of all kingdoms, which is both totally individual to each being and yet each and every one saturated with identification with Absolute Godhead. Here are some references from Aurobindo:

---

80. Emanuel, 2006, p. 154.

81. Emanuel, 2006, p. 152.

82. Aurobindo, 1996, p. 345.

"By Jivatma we mean the individual self. Essentially it is one self with all others, but in the multiplicity of the Divine it is the individual self, an individual center of the universe — and it sees everything in itself or itself in everything or both together according to its state of consciousness and point of view.[83]

And:

"The Jivatman is already one with the Divine in reality, but its spiritual demand may be for the rest of the consciousness also to realize it."[84]

Shifts in Identification: The Key is Continuity

One point that we are keen to make is that the transitions of identification that occur over the course of the path into deeper and deeper altitudes of evolutionary awakening—from the personality to soul to monadic bodyminds—should not always be understood to abstract *fully* the sense of identification from the previous altitudes. That is, we suggest that when a shift occurs from personality into soul, or from soul into the monad, it is not that identification then rests *only* at the new altitude, but rather than it can rest at the new altitude *as well as* the previous altitude(s).[85] Thus, in relation to the dance between the proximal and distal self-senses previously discussed, we still say that the 'I' of one stage is transcended and included in a higher stage so as to become the 'me'.

Prior to a monadic essence's evolutionary awakening to all the layers of identification available on the kosmic physical plane, it retains its own consciousness and self-recognition at each altitude; there is no *continuity* of self-recognition between them. That is, the monad remains forever radically Self-realized as a spark of Absolute

---

83. Aurobindo, 1970, p. 267.

84. Aurobindo, 1989, p. 20.

85. The qualifier 'can' is here used as it may be the path of a particular being to choose to abstract its embodiment, consciousness and/or identification from a particular bodymind (such as the personality bodymind, for instance, when the major locus of identification shifts into the soul) and its respective planes. In fact, as we will go on to discuss later, this way of things has for millennia been the norm owing to our collective stage of biological evolution.

Godhead subsequent to its emanative projection of a portion of itself into the soul and personality. The soul never loses its self-knowing as a pure light-being after the partial incarnation of its consciousness as a personality, and the personality remains self-conscious. What we suggest, however, is that while each of us, right here and now, exists as a radically awakened monad on the monadic planes, a self-aware soul of great wisdom on the soul planes and a self-conscious body/personality on the personality planes, it is the lack of continuity in consciousness *between all of these multidimensional altitudes of identification simultaneously* that changes as shifts in identification ever deeper again occurs.

We would thus begin to speak of beings who have engaged the shift of evolutionary awakening from the personality bodymind into the soul bodymind as simultaneously aware of their life-experience on the physical, etheric, astral and mental planes as a personality *and* as a soul on the soul planes (higher mental, buddhic, lower atmic). In relation to those masters and bodhisattvas who have completed the shift of structure-stage abiding further into the monadic bodymind also, we would see beings that are awakened on three altitudes of identification and planes simultaneously. This marks a remarkable stage of unfoldment, and relates to an evolutionary awakening interpretation of the Buddhist teaching on the possibility of a buddha residing in their dharmakaya form, their sambogakaya form, and their nirmanakaya form, and enacting variously different activities on each altitude simultaneously.

An important point in relation to the establishment of this continuity is that not every plane needs to have been fully mastered to the degree that consciousness fully directs all energy-matter on it before a being's relative identification can transcend it and shift to the next, more subtle plane. The Trans-Himalayan masters have given a very detailed explanation of this. They teach that only 2/3 of the energy-matter composing some one level of identification needs to have been brought into harmony with the higher level of identification before they can shift their identification higher onto the next, subtler level of altitude. For example, it is described in Bruce Lyon's work with Djwhal Khul that in order for a being to shift their identification from the personality to soul bodymind, which happens at what we will go on to describe as the 3rd initiation, only 2/3 of the energy-matter of the physical, emotional and mental planes needs to have been brought into harmony with the soul.

What that means is that individuals can shift from identification with the personality to soul bodymind whilst still having at least 1/3 of the interiority and energy-matter of the personality 'in the dark', as it were, to their consciousness. That means *still* governed potentially by shadow, and *still* holding and governing behavior from areas of pain, shame, guilt, fear, and anxiety that have not yet been resolved so that their energy may be released. This is one explanation of why a person can have some definitely authentic levels of access to all the wisdom, light, love, power and purpose transmission of the soul and even monadic depths of identification, and yet still sometimes be totally unconscious about all kinds of negative and destructive patterns of behaviour they are engaging in. This is equally the case for anyone engaging the Radical Awakening process also. Recognition of the Absolute does not automatically heal and bring into balance whatever shadow material one might be holding. In order for that recognition to turn into stabilized realization, it requires that we do the needed cleaning up and growing up work.

A final point to include here is that while working with the consciousness and energies of a particular altitude of identification (such as the monadic, for instance) as a temporary peak experience does not have to occur in any stage-like fashion (i.e. soul before monadic or monadic before the awareness and energies of those depths of our being residing on the kosmic planes, for instance), stabilized transitions of identification *do* occur according to uniform stages. That is, someone who is primarily identified as a personality may choose to invoke the Nondual Life-force and transformative energy of the monadic bodymind without having worked explicitly with soul-based awareness and energy first. However, as the evoked peak experiences unfold with greater regularity, the stabilized stages of identification transition they move through *will* necessarily entail a shift from the personality bodymind into the soul bodymind before the monadic, even if only for a relatively short period of the path. Indeed, this example highlights a potent approach to working with shifts in identification, as the lightning-impact and power of the monadic bodymind can greatly facilitate the opening of the personality to the soul, their integration and eventual fusion, so that the shift deeper can occur. In a planetary sense, this is why it is said in the Trans-Himalayan teachings that "Shamballa is revealed through Hierarchy but also clears the way for the emergence of Hierarchy."[86]

The Power of the Body: Kundalini

As is described above, according to the Trans-Himalayan teachings, 2/3 of the energy matter composing our physical, etheric, astral and mental bodies needs to have been brought into harmony with the soul

---

86. Lyon, 2010, p. 304

in order for the shift in identification from the personality bodymind to the soul to occur. As we will go on to describe in greater depth later, this is understood to occur at the 3rd initiation.

It is in the 4th and 5th initiations, which involve the developmental path into mastery, that beings are called to fully return to transform and transfigure the most dense, embedded, instinctual and resistant levels of their embodiment. This refers to the remaining 1/3 of their physical, etheric, emotional and mental being, so as to truly reveal as sacred their embodiment and standing on Earth.

As this process unfolds, radical awakening naturally begins to remain open and stabilized. There is a flooding of the being's embodied nature with a river of consciousness, love, and light. Deep healing is able to begin to flow into all of those as yet unredeemed aspects of their emotional nature as the downpouring soul energy brings the light-waves of consciousness to the areas of contraction trapped in the lower chakras and the physical body. This allows their energy to rise, be released and replaced by light. The soul tiers of petals within their chakra system become activated, allowing consciousness to extend deep into the etheric energy-system. Such a person's relationship with the physical body becomes one governed by love instead of unconscious, immature control. Their sexuality becomes infused with the sacred. Their use of money and other principles of survival come increasingly under the law of love. Their emotions and physical bodies become increasingly tools of planetary service, and their relationship with the Earth and the natural kingdoms transitions into one of stewardship and family under the banner of a single shared space of light rather than neglect or domination. As the soul is fused with the personality, conscious access to subtle planes in moment-to-moment waking awareness opens up along subtle correspondences to all five senses, allowing their forces and energies to be directed, manipulated, wielded, and worked with, right down into the physical plane. In the words of Djwhal Khul:

> When the life of the personality is carried up into Heaven, and the life of the soul comes down on to earth, there is the place of meeting, and there the work of transcendental magic becomes possible.[87]

When identification is then increasingly deepened into stabilization in the monadic bodymind, which marks a highly advanced stage of unfoldment,[88] and brought into relationship with our incarnated,

---

87. Bailey, 1934, p. 250.

88. We will be exploring such advanced stages of development in detail in Chapter 16.

emotional, etheric and physical nature, an extraordinary expression of embodiment is able to flourish. The release and transmission of divine Power, Will, Purpose and Life in the heights begins to allow a penetration into and the revelation of sacred power in the depths. According to the Trans-Himalayan teachings, while the consciousness and energy of the soul is able to penetrate and bring love and light to fully solarize the mind and emotional natures, it takes the presence and power of the monad to penetrate all the way into and access the innate divinity of the physical body. This brings the most base and instinctual drives of our nature toward self-preservation and survival at any cost that reside there under the service of God. It is at this depth of the body, deep in the base chakra, that we find the primal *will-to-live*.

As was pointed to earlier, in the Trans-Himalayan teachings, just as the monadic essence is understood to be the emanated Self and Life-force of the One Universal Monad, which is itself the dynamic Life-force of Absolute Being, so also are the soul and personality the emanated Life of the monadic essence. The point of anchorage for that emanated Life in the soul, which acts as its animating base, is its central jewel, which we described earlier. The point of anchorage for the emanated Life of the monad in the personality bodymind is this reservoir of the will-to-live in the base chakra—the kundalini energy, as it is known in some Eastern traditions.

Kundalini is the presence of the dynamic Life of Awakened Awareness in the base chakra and energy systems of all human beings. It is confined there as a result of their self-contraction into and misidentification with the personality bodymind. Really, it is the power or will-to-live of Absolute Spirit, but while it is confined in the base chakra of a human being by their self-contraction, it expresses as their personal will-to-live.

The power of this will-to-live is extraordinary, and this presence of monadic power in the depths of our being is the counterpoint to the monadic power found in the heights. This normally unconscious aspect of our nature has the capacity to totally overpower the soul's morality of love or a spiritual practitioner's apparent stabilization of radical awakening in the most stark and savage times. It is able to demonstrate outstanding bravery and courage with no consideration for its own survival, which may be far beyond the capacity of the sensitive soul. According to the Trans-Himalayan teachings, buried in the earth of this level of our being is that pure instinctual drive to survive in the dark regions of this world never yet touched by light, so that the physical body might live long enough for our radically awake monadic God-self to arrive eventually. This is the power, rooted in the equanimity of absorption in unbounded Awakened Awareness, to act, transmit and ground kosmic Purpose in this world with a might that far transcends that mustered by the forces of opposition it will necessarily come up against. This is why it is said in the Trans-Himalayan teachings that:

> *The will ever implements the purpose.* The repository of the will aspect of man's innate divinity is to be found at the base of the spine; this can only function correctly and be the agent of the divine will after the third initiation.[89]

It is this will-to-live found in the base chakra that becomes fuel to be consumed in the heart of the now monadically identified master when, from a space of total non-judgement and equanimity, they return to make sacred and unleash the power of their embodiment on Earth. This is the kundalini energy that when freed from its confinement in the base chakra through the release of self-contraction into radical wakefulness comes online as the Parashakti, or Supreme Life-force of Absolute Reality, to express itself selflessly and willfully.

As this unfolds, the physical body is transfigured into a temple of Eden, and the fusion of radical wakefulness, monadic Life and physical embodiment is grounded so fully that the cells of the body begin to literally shine with light, as has been demonstrated by some of the greatest masters and teachers of the past. This allows the monadic Life-essence at the core of every cell and atom of the physical body, the product of Absolute Reality and its Life-force expression's apparent self-contraction into the form of a personal bodymind, to become activated so as to saturate every muscle, nerve fiber, bone, and organ of the body in unbroken bliss. It allows the black hole capacities of the monadic bodymind to bend and warp space and time in conformity to divine Purpose, to become the property of a master on the physical plane. Then, the powerhouse of their very physical presence becomes an innately purifying, healing, empowering, and stimulating force within their environment in an incredibly potent way.

Eventually, as 2/3 of the entire kosmic physical plane—which amounts to the whole of the lower five planes (physical, astral, mental, buddhic, atmic)—is transmuted, transformed, and transfigured, the radically awake monadic essence chooses whether to transcend or stay again, but now on a kosmic scale and in relation to the *entire kosmic physical plane.* One possibility is that it may now release all the layers of identification it has been working through and choose to enter into new spheres of service on higher kosmic planes. Alternatively, it can choose to stay and take up its place in the community of buddhas who together form the collective community of Shamballa on the highest two sub-planes of the kosmic physical. It is these beings who together hold identification with and transmission of the divine Purpose of

---

89. Bailey, 1960, p. 714.

Earth in the kosmos, and the Power and Will required to implement it. This leads us nicely into our next chapter.

Chapter 11
Relationship
Jon

The second line of development within the Evolutionary Awakening vector relates to the cultivation and experience of relationship across multiple dimensions. From each of the levels of identification for the monadic essence we spoke of in the previous chapter (the personality, soul, and monadic bodyminds), relationship can unfold with other beings that may be residing on denser planes, the same planes, or subtler planes. In terms of the service each being renders throughout the great chain of the hierarchy extending through the kosmos, the hand of relationship, empowerment and service may be held out to all beings. Some of those will be more developed, some less developed, and some at the same level of evolutionary altitude.[90]

A Note on Development

Before we go any further, it serves to acknowledge that according to the model of the path that we are exploring in this book, assessing the development of a particular being is not something that can be measured according to a single metric. Rather, this needs to take specific context, content, and capacities into consideration. For instance, a being might have a solidly stable degree of radical awakening and yet may not have unfolded the identification line of their evolutionary being deeper than

---

90. It might seem here that a correspondence is being drawn between a being's plane of abiding and their level of development. Sometimes it is the case that these two reflect each other, but also sometimes not. Thus generally, we like to keep them differentiated.

the mental aspect of their personality. They equally might not have opened up multiple dimensions of conscious relationship or plane access, though they may be operating through structures of intelligence that are rare in their sophistication among the majority of humanity. Alternatively, a person may have access in their identification to the soul bodymind, may have opened up conscious awareness of relationships with liberated masters on the soul planes, and yet may have not yet ever had a full experience of radical awakening. While they may have unfolded lines of soul-based intelligence (e.g. intuitive mind, illumined mind), they might not have unfolded lines of mental, emotional, etheric or physical plane intelligence to nearly the same degree.

This calls us to be cautious before we try to judge development according to only those lines of unfoldment with which we might be most familiar. As we are hoping to show in this book, the lines of radical and evolutionary awakening are diverse, with the spiritual traditions displaying extraordinary levels of wisdom in one or two of them, but very rarely more than that. It is our sense that only through the recognition of each line as a unique path in itself that a true, multidimensionally oriented universal spirituality may emerge. Rather than being an immature homogenization that dissolves the unique features of each tradition, we envision such a universal spirituality will be able to recognise and honor their particular line of gifts. It will be able to do this whilst still understanding all as branches of the one planetary lineage of awakening—the Great Human Tradition.

The Multiple Dimensions of Relationship

As previously said, relationship can grow and be cultivated throughout the various sub-vibrational frequencies of the kosmic physical plane (physical, astral, mental, buddhic, atmic, monadic, logoic), and beyond into more subtle kosmic planes (kosmic astral, kosmic mental, kosmic buddhic…), with the different threads of relational connection being woven between beings on the same, higher, or lower planes relative to each other. Ultimately, it is through the communities of lives that reside on the different vibrational planes standing hand in hand with each other in the shared transmission of Light, Love, Power, and Awakening that the unfoldment of the kosmic Purpose of the Earth may be empowered, consciousness expanded, and all forms revealed as divine.

In this section we will explore the types of relationship that can be cultivated both vertically (with beings on other planes, both higher and lower) and horizontally (with beings on the same plane) from the personality, soul, and monadic altitudes of identification. It should be noted that this use of the terms horizontal and vertical does not equate to our use of them in relation to radical and evolutionary awakening. Rather, as they are used in this chapter, they relate to different orientations of relationship cultivation that can be unfolded within the evolutionary awakening vector.

Relationship at the Personality Altitude: Horizontal

Horizontally, relationship at the personality altitude (physical, emotional and mental) is one of the richest, most fulfilling aspects of human life. Examples here are those personal friendships with those we grow up with and come into contact with in life to share smiles, laughter, fun, and adventure. Or the collaborative relationships we experience in our work lives with those by whom we are stimulated and inspired. Horizontal relationship at the personality level expresses also in the sacred sisterhood or brotherhood of the path within which it is our privilege to see and be seen by each other in so many more ways than we may experience in the more conventional arenas of life. It can express in terms of romantic relationships of sacred physical, emotional, intellectual and soulful intimacy, or any combination of these. Such relationships stand as what must be some of the most hallowed currents of the life river.

On physical levels, we can speak of the cultivation of sacred relationship between physical bodies, whether in relation to sexual intimacy, or perhaps with those with whom we might engage in bodywork processes (e.g. massage, breath work, Bioenergetics, Hatha Yoga, Tai Chi, Qi Gong, dance, physiotherapy, acupuncture) so as to heal and/or promote the greatest health in the physical nature. Two examples of such approaches point us to the spiritual tradition of tantra and sacred sexuality. Here it is our embodiment and sexuality that serve as the basis of exploration into the esoteric mysteries of the physical body, the masculine and feminine faces of divinity, the Earth and physical kosmos.

In relation to horizontal relationship at the emotional altitude, here we can speak of those professional or personal relationships in which space is held for emotional work, emotional healing and feeling-centered relating, as well as for fun, humour, and the pleasure of life-enjoyment.

Horizontal relationship on the mental plane expresses in terms of working with others in the realm of ideas and concepts, as well as developing relationships with those ideas and concepts too. It involves the capacity to midwife intuition and sparks of inspiration into unpacked concepts and structures of thought, and then to be able to communicate them in a creative, skilful, delicate and clear way. This might express in scenarios ranging from professional circles to therapeutic scenes to social spaces in which we play with others in the field of ideas for creative stimulation.

Through all of these altitudes and those whose discussion is still to come, the consideration of sacred relationship with another being asks us: "How many levels of union are we experiencing in any of our relationships?" Just as it is possible to come into physical union with another human being, so also is it possible to come into emotional union, mental union, or soul union, and to recognize already present monadic union. Often personal relationships will only be rooted in one or two of these. While that may be deeply nurturing and stimulating for those involved, some of the most inspiring and dynamically creative relationships are those in which we experience union with another human being on multiple levels.

In our writing process for this book, Dustin and I have definitely experienced our always-already recognition of union and deep love for each other as brothers on the plane of soul extending to profound and sophisticated degrees onto the planes of mind and emotion. The fact that relationships that extend union across multiple planes are normally numerically fewer than others calls us to recognize just how sacred they are. This illustrates what a glorious opportunity it is to share the revelations of some of the richest mysteries of life with another reflection of the Godhead on multiple levels.

Relationship at the Personality Altitude: Vertical

Vertical relationship at the personality altitude involves the extension of lines of connection from one of the personal planes (physical, etheric, emotional, and mental) to beings or aspects of their nature operating on other planes. Some examples of this type of relationship in the esoteric arena are psychics or mediums, those who practice alternative forms of healing, and shamans.

Those we term psychics, channels, and mediums, when genuine,[91] can be those who have established relational contact with non-physical beings dwelling on the mental, emotional, or etheric planes, and into the subterranean layers of the Earth. This also goes for some of the work of shamans. Such contacted beings may be human souls out of

---

91. This qualifier is important. Often, a mature eye can recognise that the source of a psychic, channel or medium's communication is very likely an unconscious aspect of their own psyche, or is drawn from their exploitation of basic commonalities of and assumptions about human nature.

incarnation, perhaps, or beings such as devas. Occasionally the source of transmission is from higher planes too.

Relationship at the Soul Altitude: Horizontal

Horizontal relationship at the soul altitude occurs between souls residing within overlapping and interpenetrating fields of wisdom-consciousness and love-energy. As many of the esoteric teachings present it, and as is confirmed and supported in great detail through the research of such trans-personal hypnotherapists as Michael Newton, Joel Whitton and Joe Fisher, souls reside in great groups or constellations of light on the soul planes. The Trans-Himalayan teaching understands the major grouping of souls on the inner planes is *hierarchically* organized on the basis of the inclusivity and wisdom of their consciousness and energetic presence, and differentiated *typologically* according to the pattern of the Seven Rays. The hierarchical arrangement includes all human and post-human souls within the aura of the Earth. This extends all the way from the very youngest souls to those whose lotus bodies have unfolded perhaps one or two of the knowledge petals, to those who are unfolding the love petals, and more advanced souls also. These latter might be those who are unfolding the sacrifice petals in the stabilization of radical awakening, or the community of soul elders and masters who stand at its summit of the soul planes.

In terms of the typological differentiation of these soul groupings across the Seven Rays, it serves us to note that that the Seven Rays are understood as the seven major energies of the subtle evolutionary kosmos. In the Trans-Himalayan teachings, they are described as pouring through and coloring every being's chakra system, to greater or lesser degrees, depending on the purity of transmission capacity. This extends from that of an animal or human being to a kosmic Logos manifesting through a planet, sun or galaxy.

The Seven Rays form the major typological system of the Trans-Himalayan teachings, and one that is multidimensionally oriented. This means that every depth of identification,[92] every major field of human activity, each of the major soul and monadic altitude groups, as well as the planes themselves, are understood to be typologically vibrating predominantly to one of these seven energies.

---

92. While the personality and soul depths of identification are understood to vibrate potentially on any of the Seven Rays, the monadic essence depth of our nature will vibrate only on one of the first three.

Ray	Human fields of activity	Soul Group	Major plane of resonance
First Ray of Will and Power	Politics, Law	1st Ray Ashram	Logoic
Second Ray of Love-Wisdom	Spirituality, Healing, Education	2nd Ray Ashram	Monadic
Third Ray of Creative Intelligence	Economics, Business, Philosophy	3rd Ray Ashram	Atmic
Fourth Ray of Harmony through Conflict	Art, Music, Drama, Culture, Media	4th Ray Ashram	Buddhic
Fifth Ray of Concrete Science	Science, Technology	5th Ray Ashram	Mental
Sixth Ray of Devotion and Idealism	Religion, Mysticism	6th Ray Ashram	Astral
Seventh Ray of Embodiment, Magic and Ceremony	Ecology, Environment, Earth-centred orientations, Indigenous Wisdom	7th Ray Ashram	Physical

Table 2. Table showing the correspondences between the Seven Rays and the fields of human activity, soul groups and planes through and on which they express.

On the soul planes, we would therefore speak of seven major soul groups, or subtle ashrams, together comprising the great community of all Earth souls, with each ashram composed of souls vibrating primarily to the energetic frequency of one of the Seven Rays. Presiding over these subtle ashrams are the masters, bodhisattvas and elders in spirit who together form what the Trans-Himalayan teachings call the Hierarchy—the planetary meta-sangha.

Hierarchy: The Planetary Meta-Sangha

The existence of greatly enlightened beings working on the soul planes, who remain within the aura of the Earth to stimulate and empower planetary awakening and evolution, is near unanimously attested to both in the traditions.[93] It is also testified to in the experience

---

93. Many of the most revered spiritual texts within the traditions flow from inner contact with such beings. The Longchen Nyingthig teachings in Dzogchen, revealed to Jigme Lingpa, as well as the teachings revealed by Dudjom Lingpa, are examples of this. In fact, the Nyingma terma, or revealed treasure tradition, continues to serve as

of more and more human beings and groups. This is the 'Rijjall ul Ghaib', or the 'invisible world government', of the Sufis; the Ancestors of the Shamanic and Earth-centered traditions; the Gyanganj or Siddhashrama of the Hindus; "the spirits of just men made perfect" of the Bible (Hebrews 12: 22-23); the liberated buddhas and bodhisattvas of the Buddhists; the 'Tzadikim Nistarim' or 'Hidden Righteous Ones' of mystical Judaism; the 'Xian', or Immortals of the Taoist; and the Hierarchy of the Trans-Himalayan teachings.

From an Integral perspective, the Hierarchy can be understood as a LL collective culture on the soul planes rooted in the shared ethic and worldview of soul—lighted mind, pure love-wisdom, and enlightened will in planetary service—that has its own planetary energy field(s) in the LR. As Djwhal Khul describes it, this planetary meta-sangha of Hierarchy and its soul ashrams are characterized by *group consciousness*. As human beings stabilize both radical awakening and their center of gravity in the soul bodymind, they establish continuity of consciousness across the altitudes of the kosmic physical plane. This allows them to remember in waking awareness the altitude of identification, the relationships, the plane experience and lines of soul-based intelligence they are *already functioning through* within the planetary meta-sangha of Hierarchy. This now allows them to expand consciously and develop these lines as well as their planetary service within that collective culture of the soul planes, working within one of the soul ashrams that compose it.

As was mentioned in the previous chapter and will be explored more fully in the next chapter, the soul planes are those upon which the Subtle state of Absolute Reality shines forth most evidently. Therefore, beings operating on the soul planes work within the fields of subtle light, sound, and rays. Unlike in the Causal state, where consciousness is formless, and the Gross where consciousness is fixed, consciousness arising in the Subtle state and the planes pervaded by it is permeable in its boundaries. It is this that allows Hierarchy as a whole to function as a vast field of shared, interconnected consciousness, and that affords all within the ashrams the unified response to the planetary Will, Purpose and Life transmitted into it by the most advanced masters from Shamballa, a monadic plane community of beings that we will go on to explore in the next section. It is abidance in this field of

---

a vehicle for such subtly sourced teachings. Examples of the same in other traditions abound, as in the contemporary experience of modern day teachers such as Thomas Hubl and Anadi also. This is the manner in which the Trans-Himalayan teachings have been transmitted, via subtle level rapport from such masters as Djwhal Khul, Morya and Rakoczi to Alice Bailey, Bruce Lyon, Helena Roerich, Helena Blavatsky and Lucille Cedercrans.

vast shared consciousness that allows total telepathic intimacy across space and time, full psychic connection, loving sight of the whole of humanity and the other kingdoms, and the shared registration of planetary consciousness and form as it exists on kosmic planes. This is the basis of the process of group initiation, or the movement through advanced initiations of unfoldment *as a single WE*, which characterizes Hierarchy and which opportunity is increasingly available to human groups seeking to pioneer new forms of collective evolution.

These soul ashrams of Hierarchy exist like galaxies on the soul planes, with a myriad of sun-souls circling a central dark Presence—that of a radically awake, monadically identified master at its center. Taking a glimpse at the way we could understand the development exemplified by these beings gives us a nice illustration of how the important differentiation between the radical and evolutionary vectors, and the lines of the latter, can play out in advanced stages of development. The masters who stand as the heads of the soul ashrams can be understood as radically awake beings whose identification is stabilized in the monadic bodymind, but who choose to serve through their reception to and transmission of the Will, Purpose and Life poured forth from Shamballa into a field of souls on the soul planes. Indeed, at their stage of highly advanced development, the sphere of their enlightened will, love and energetic presence is so large, and their mastery of energy manipulation so sophisticated, that in the Trans-Himalayan teachings it is understood that the soul ashram over which they preside actually becomes the soul bodymind that they *incarnate through* on the soul planes. That is, the number of beings their soul bodyminds have begun to be able not just to embrace in love, light and enlightened will, but to hold an energetic field of empowered development for, has expanded so widely that they are able to *incarnate through an entire ashram*.

To take this point further, at the highest level of the planetary meta-sangha, the Hierarchy as a whole is actually understood in the Trans-Himalayan teachings to be the single soul bodymind through which the vastly realized being, Maitreya, is now incarnating. Maitreya is the being prophesized in Buddhist teaching as the next buddha that is to incarnate on Earth after Gautama Buddha some 2,500 years ago. In the Trans-Himalayan teaching, Maitreya is actually understood as the same figure as the Christ, though the latter term is used to refer to a position in Hierarchy—that of the World Teacher for any particular major time cycle (such as the approximately 2,250 years that comprise the astrological ages). The World Teacher is described as the being who takes responsibility for initiating the major spiritual and religious movements for any major cycle of time that may support humanity's growth and awakening, such as over the last and current astrological ages of Pisces and Aquarius respectively. In this connection, it is interesting to consider that just as in the West the being that initiated the largest extant spiritual movement is the Christ, in Tibet the legend runs that it was from Maitreya that the Indian sage, Asanga, received the transmission that formed some of the core teachings of the Buddha-Nature, or *tathāgatagarbha sutras* (principally, the Uttaratantra Shastra). These focus almost exclusively on the true nature of every being as Awakened Awareness, and inspired the third turning of the Buddhist Wheel of Dharma that emerged eventually as Vajrayana Buddhism. It is inspiring to contemplate the possibility that a single being's transmission and life could have been responsible for such profound spiritual movements on Earth as Christianity as a whole and the foundational radical awakening teaching of Tibetan Buddhism. Indeed, this perhaps points to what it truly means to stand as the World Teacher for any particular astrological age, or series of ages.

Of course it is important when speaking about such beings to not reify identity—i.e. that of this enormously evolved being known as Christ, or Maitreya, as being eternal, real, independently existing, etc.—but rather to be open to the *radical* perspective also. Bruce Lyon and Djwhal Khul do a nice job of keeping us on track here:

> Whatever name we use for the Christ or Maitreya for example, there is an underlying reality that transcends identity. It is not an identity who has become the Christ but a principle that has let go of its lesser identification. This is not possible to explain to those beneath the Third Degree because it causes the identity of the reader to confront its own mortality.[94]

This is a fundamentally important point for this entire book. Masters, buddhas, planetary, solar, or galactic Logoi; all of these beings are no more than Awakened Awareness arising in radiantly awake energy, in the apparent form of such a being. In truth, as Bruce Lyon and Djwhal Khul say, the development of such a being is not so much the ascent of a self, but rather the continued declothing of Reality's arising in a particular wave of its emanation.

Returning now to our consideration of how for advanced masters such as Maitreya, also known as the Christ, whole ashrams (or the Hierarchy as a whole in the latter case) embody their soul bodyminds, here is the Master Rakoczi:

> All members of an Ashram, from the Master to the member within the sphere of influence, partake of one state of consciousness, insofar as their awareness permits.

---

94. Lyon, 2010, p. 220.

Thus, in one sense, every member of an Ashram is part of the Consciousness of a Master.

Actually, that magnetic field of consciousness which lies between the positive pole of Spirit and the negative pole of Matter, is the Soul of the Master. I would have you consider this. A Master functions in the Monadic level as a unit of life, and what I shall refer to you for your benefit as consciousness, but this is a misnomer. It is individuality identification within the Christ. The consciousness of the Ashram, shared by all the membership of the Ashram according to the ability of each to share in it, is the Soul of the Master. The body itself is the Etheric Light Body which emanates from the Soul Life of the Ashram, from Buddhic Spheres, and constitutes its 'ring-pass-not', so that the Ashram, in the first sense, is maintained above the three worlds of human endeavour.[95]

Like a supermassive black hole at the core of a galaxy, it is the master at the core of an ashram whose awakening, presence, consciousness and energetic field is so dynamic and magnetic that it both holds the whole soul-field in a state of coherence, as well as incarnates their scale of kosmic realisation and identification with planetary Purpose through it. In doing so, all souls within their great soul bodymind that is the ashram are themselves stimulated and empowered to grow and unfold over centuries and millennia, into mastery. Necessarily also, to extend the supermassive black hole correspondence, the dynamic Presence of the master at the center continuously pulls all the orbiting sun-souls within their ashramic soul bodymind closer into the core. In this process, the spiral-cyclic pattern of movement symbolizes the ages-long journey of a soul within the great School for Souls that is the Earth, as it orbits ever more closely the omnicentered field of Reality itself.

From an Integral perspective, all of this calls us to vision an entirely new AQAL matrix for such beings, and this can be seen below.

---

95. Cedercrans, 2004, p. 34.

Individual

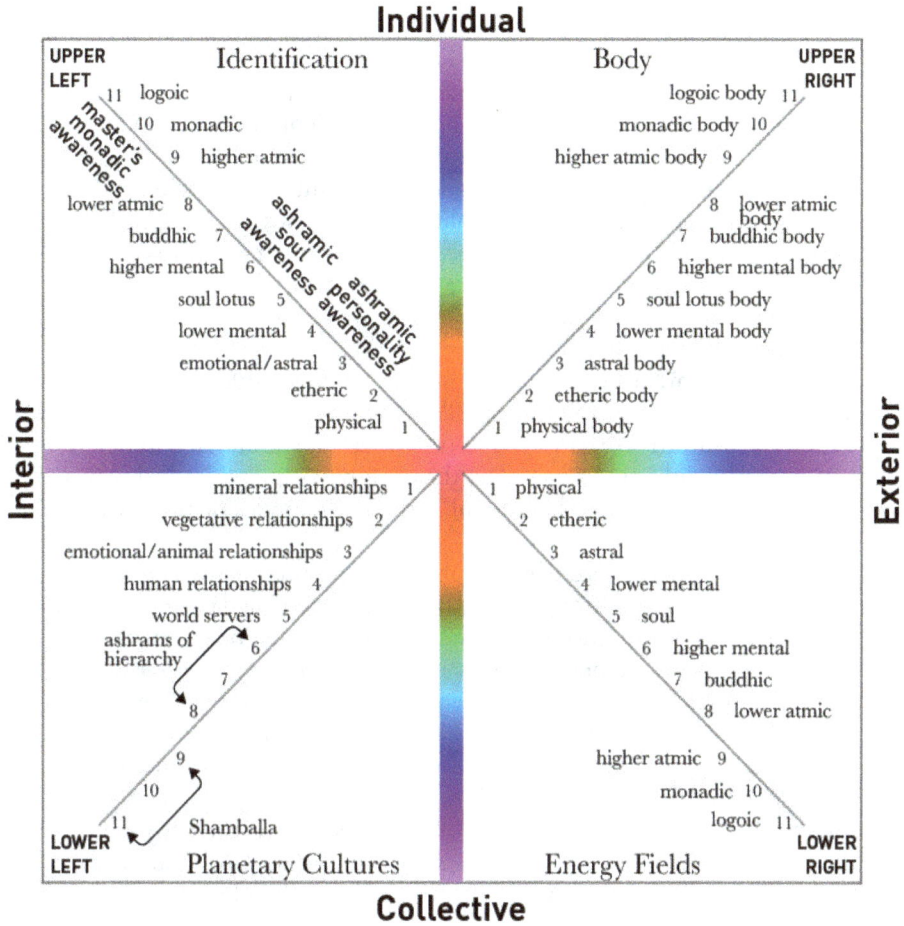

UPPER LEFT	Identification	Body	UPPER RIGHT

Identification (Upper Left)
- 11 logoic
- 10 monadic
- 9 higher atmic
- 8 lower atmic
- 7 buddhic
- 6 higher mental
- 5 soul lotus
- 4 lower mental
- 3 emotional/astral
- 2 etheric
- 1 physical

master's monadic awareness
ashramic soul awareness
ashramic personality awareness

Body (Upper Right)
- logoic body 11
- monadic body 10
- higher atmic body 9
- 8 lower atmic body
- 7 buddhic body
- 6 higher mental body
- 5 soul lotus body
- 4 lower mental body
- 3 astral body
- 2 etheric body
- 1 physical body

Planetary Cultures (Lower Left)
- mineral relationships 1
- vegetative relationships 2
- emotional/animal relationships 3
- human relationships 4
- world servers 5
- ashrams of hierarchy 6
- 7
- 8
- 9
- 10
- 11 Shamballa

Energy Fields (Lower Right)
- 1 physical
- 2 etheric
- 3 astral
- 4 lower mental
- 5 soul
- 6 higher mental
- 7 buddhic
- 8 lower atmic
- higher atmic 9
- monadic 10
- logoic 11

LOWER LEFT	Planetary Cultures	Energy Fields	LOWER RIGHT

Interior — **Exterior**

Collective

Figure 19: The multi-dimensional Four Quadrants of a master incarnating through an ashram.

Exploring this, we can see that the LR plane fields stay the same as for any other monadic essence incarnating through the kosmic physical plane. There is equally not much obvious difference in the LL quadrant, as the collective cultures and worldviews that such masters have access to at the monadic, soul, and personality altitude remain as those held by the planetary communities of Shamballa, Hierarchy, and humanity as a whole. The real changes have occurred in the upper quadrants, where the constituent holons of what we would normally classify as collective fields now have the opportunity to participate in the individual identity (UL) and bodies (UR) of the master. These collective fields should be understood as constituted similarly to the collective fields of atoms, molecules, cells, biological organs, etc., that compose the UR physical body, but at a higher scale of inclusivity.

So, returning to the AQAL matrix above, in the UL individual interior quadrant, we can see that the master, identified in the monadic bodymind, has also available a level of soul consciousness in which all souls within that ashram have the opportunity to participate. And the master has a level of personality consciousness in which all the personality consciousness of those of its members incarnating on the mental, emotional and physical planes have the opportunity to participate. In terms of the UR quadrant, we can see that the master's monadic identification is still anchored in their monadic body, but their soul body is now a light body of higher mental, buddhic and lower atmic energy-matter containing all the smaller soul bodies of those within the ashram. We can see that their mental, emotional and etheric bodies are sheaths that emerge as they are participated in and woven by the totality of mental, emotional and etheric bodies of those within the ashram.

It may be asked here why the physical body manifestation of such masters incarnating through ashrams is not seen on Earth yet. The reason for this is tied into our current collective stage of evolution and awakening on the planet. *The ashrams do not yet have physical bodies*, and it might equally be pointed out that their etheric and perhaps even astral sheaths still remain nebulous. The masters incarnating through these ashrams, who at their stage of development are now only spaces through which the energy of Absolute Reality may freely express, are engaged in an ancient project of progressively externalizing onto the outer planes. Indeed, the very same can be said for the Christ's incarnation through the planetary Hierarchy as a whole.

This is what is described in the Trans-Himalayan teachings as the Externalization of the Hierarchy and the Reappearance of the Christ. In order for these evolutionary unfoldments to be carried through into fullness on the mental, emotional, etheric and physical planes, it will require the necessary complexity of individual and collective consciousness, behavior, culture and society to support it.

Our current era is one of great crisis economically, environmentally, culturally and socially, and yet there are more and more signs that in various locations the evolution of the new forms of consciousness, behavior, culture and societal structure are emerging. In the language of the Trans-Himalayan teachings, these may eventually be able to support this Externalization of the Hierarchy and Reappearance of the Christ. More universally, we can understand this as the full incarnation of Love onto Earth, and the revelation of planetary Purpose. As this occurs, in the Trans-Himalayan teachings it is prophesized that humanity and Hierarchy will be fused; the ashrams of the planetary meta-sangha will eventually be anchored in physical locations on Earth through which the Purpose of our planet in the kosmos may be expressed. It is taught that to support this process, the masters and high initiates of the planetary meta-sangha will externalize and walk again openly on Earth, and the collective culture and civilization of the planet will be rooted in the pure ethic of the soul, which is love.

The Human Perspective

In terms of the horizontal forms of relationship that characterize human beings such as ourselves at the soul altitude, there are two areas to explore here. The first relates to guides on the soul planes, and the second relates to relationships shared within soul groups.

In connection with guides, Michael Newton's research has documented the corroborative reports of people who have been hypnotherapeutically regressed into the soul altitudes of their being. Such people describe how all souls of every stage of development within the soul planes share relationships with guides of greater development. This includes the guides themselves with progressively advanced initiate-souls and masters of Hierarchy. Here is Michael Newton:

> I have never worked with a subject in trance who did not have a personal guide. Some guides are more in evidence than others during hypnosis sessions. It is my custom to ask subjects if they see feel a discarnate presence in the room. If they do, this third party is usually a protective guide. Often, a client will sense the presence of a discarnate figure before visualizing a face or hearing a voice. People who meditate a great deal are naturally more familiar with these visions than someone who never called upon his or her guide. The recognition of these spiritual teachers brings people into the company of a warm, loving creative power. Through our guides, we become more acutely aware of the continuity of life and our identity as a soul. Guides are figures of grace in our existence because they are part of the fulfillment of our destiny. Guides are complex entities, especially when they are Master guides. The awareness level of the soul determines to some extent the degree of advancement of the guide assigned to them. In fact, the maturity of a particular guide also has a bearing on whether these teachers have only one student or many under their

direction. Guides at the senior level of ability and above usually work with an entire group of souls in the spirit world [the soul planes] and on earth. These guides have other entities who assist them. From what I can see, every soul group usually has one or more rather new teachers in training. As a result, some people may have more than one guide helping them.[96]

In relation to such soul groups, these are our soul family—the group of souls with whom we have journeyed for perhaps centuries or millennia in and out of incarnation with, and for whom the soul depth of our hearts vibrates always with profound affection and affinity. Their spheres of relationship are large. In Michael Newton's research, individuals in a state of transpersonal hypnosis describe 'primary' soul groups as made up of anywhere between three to twenty-five souls, with many of these primary soul groups composing a much larger 'secondary' soul group of about 1000 souls.

The primary soul groups are those particular constellations of deep soul connection that carry on over lifetimes. These primary soul connections exist with those for whom we feel an *immediate recognition* upon meeting or knowing, one that is instinctual and intuitive yet not readily understandable by the mind. It is as if from behind and beyond our little memory from just this one life there resides a whole body of history in shared journeying and deep feeling. Again, with recognition that such intuitions may have their basis in reality, we can understand the lack of full consciousness concerning such connections to be related to the lack of established continuity in consciousness between the soul and personality altitudes of identification. We can understand that as this changes over the course of our evolutionary growth, the sacred details of the trans-incarnational fellowship we share with those with whom we stand closest become clear.

For some, such relationships will be found in their biological family. For others, they will express in the field of friendship or romance. Such relationships based on shared soul connection may work out in our professional life. Whilst identification resides in the personality bodymind (physical, emotional, and mental), these soul-based connections will most often be found in the individual's own sphere of personal interactions, and will have most impact there. When the center of gravity is shifting into the soul, the connections will often begin to express in transpersonal spheres that are rooted in world service, consciousness and love (though may still be found in biological family and friends).

---

96. Newton, 1994, p. 107.

In this process, and as the influence of the soul grows in the life of an individual, their understanding of the piece of world service it is their unique destiny to give progressively reveals itself. As this transpires, they will often move more and more purposefully into that field of human activity that corresponds to their soul ray (see the ray table above). In doing so, they progressively come into relationship with those other souls with whom they stand in one of the seven ray ashrams on the soul planes, and the various patterns of karma and destiny that they collectively share will move their lives like waves upon the ocean.

As identification shifts into the monadic altitude and radical awakening is opened up as the Indivisible Reality of Being through which the transmission of Purpose and kosmic Power flows, there will arise relationships expressive of Reality in communion with Itself, of the primordial Purity of the All, and the enjoyment of Earth as Eden. In the words of Djwhal Khul:

> Once LIFE becomes a reality within the 'identity' of the initiate then lesser identifications fade. He or she may still form an essential part of groups, organisations, nations but their anchor will not be 'within' their outer group but with the essential Life expressing through it. They become ambassadors of the One Life and through their identification with it which increases as the higher initiations are taken, they induce its realisation in others. All lesser fires are contained within the one 'synthetic' fire which underlies the cosmic physical plane, and whenever this fire of synthesis is present even in the tiniest of 'doses' it has a homoeopathic effect upon its surroundings. Such workers must remain 'free' to follow the call of spirit when it comes. They are leaves which are moved by the wind which "bloweth where it listeth" (John 3:8). They must learn to respond to that inner call, the straight knowing which flows down the fiery current of their anchored antahkarana [the continuity of consciousness established between the personality, soul, and monadic altitudes of identification]—and in such a way they will widen this channel of fire. They are no longer held by group affiliations, loyalties or commitments but must hold their allegiance to something deeper that beats at the heart of all groups. Their comings and goings will then be 'unreasonable' to the consciousness operating within such lesser groupings, but this very unreasonableness will be a signpost to that which lies beyond reason—the spiritual instinct which acts without thought because it is not inherently separate from the life which motivates it.

There is in fact no separate identity to 'think' about what 'it' should do.[97]

Karma and Destiny

According to the Trans-Himalayan teachings that have come through the collaboration between Bruce Lyon and Djwhal Khul, both personal and soul-based relationships can be anchored in either the Law of Karma or the Law of Destiny. When rooted in the Law of Karma, it is the past that draws and holds us in relationship with each other. This often found to be the case in friendships, romantic relationships, and particularly in the case of biological family relationships over generations. These latter are often cases where the DNA inherited by younger generations serves as one the most fundamental reservoirs of karmic challenge and opportunity to face and work through whatever might be called for as a result of that which has occurred in previous incarnations. When a soul starts to come under the influence of the Will-force and Purpose identification of the monad, we start to become drawn and held in relationship by the Law of Destiny. Then we could say it is the *ageless future* that acts as the magnet for relationship—ageless because this future is and always has been one with the intent pouring through Absolute Reality itself. This shift so as to come under the influence of the Law of Destiny allows entirely new configurations of relationship that are fresh, and not necessarily rooted in any previous karmic ties, to be born.

As the souls of a particular constellational group within one of the seven great ashrams of Hierarchy mature, one of the things that starts to happen is that the bonds that hold them together begin to vibrate with what could be called *post-autonomous communion*. At this stage, each soul has awoken their capacity to stand alone as a living photon of kosmic light on Earth, and from that position of spiritual and personal autonomy can now choose to enter spaces of communion with others that are all the more powerful. This is because each has learnt to stand autonomously and move freely from their own awakening to Source.

It is in this intimacy of personality, soul, and monad—coming increasingly under the Law of Destiny—that Dustin and I have experienced the honor to share in as our brotherhood and work in the world. It is this drive of Destiny that we increasingly share with sisters and brothers around the world, of many backgrounds, traditions, fields and nations. As was mentioned in our *Note from the Co-Authors*

---

97. Lyon, 2005, p. 290-291.

at the beginning of this book, within the first minutes of meeting we recognized each other as here to serve the same purpose. Since then the relationship, which we know and experience as rooted in the very depths of our souls, has continued to mature. It is one that we increasingly experience to fly on the wings of a love and brotherhood so full that we are able to show up for, hold ourselved accountable to, and support each other more and more powerfully in our awakening, integration, and presence for the one work of which we are a part.

Inspiringly, we know that our experience in this is not unique, and that all over the globe maturing soul sisters and brothers are finding each other. When monadic Will and Destiny starts to come online, it has the capacity to slice through the apparent limitations of space and time, thus allowing the connections of the ageless future, which are so crucial for the planet at this time of great crisis, to be made.

Relationship at the Soul Altitude: Vertical

Vertical relationship from a center of gravity in the soul has the capacity to extend 'downwards' into evolutionary life-spheres of humanity, the animal, vegetable and mineral kingdoms on the mental, emotional, etheric and physical planes. It has the capacity to extend outwards into the planetary soul-plane community of Hierarchy on the higher mental, buddhic, lower atmic planes. It can extend 'upwards' toward Shamballa on the higher atmic, monadic and logoic planes. This is the profound capacity of the soul to be able to bring every depth of the self and kosmos into relationship.

Contact

If we consider the extension of relationship out into Hierarchy, particularly into its most advanced spheres of being, this takes us back to the point made in the previous discussion on the masters. One area it is of profit for us to explore here relates to the contact that may occur between an individual and one of these masters, who together stand as the consciousness-elders of our planetary life and who are engaged in kosmic scales of service from within the aura of the Earth.

One point of teaching that makes the Trans-Himalayan teachings unique is the extensive body of transmission on this topic that they both embody and contain. Despite the fact that some of the masters are definitely understood to occupy and work through physical bodies,[98] for the most part the method of transmission engaged in Trans-Himalayan teachings has been consciousness-to-consciousness from remote distances. This was the case when Helena Blavatsky worked with the Masters Morya, Koot Humi and Djwhal Khul; when Alice Bailey worked with Djwhal Khul; when Helena Roerich worked with Morya; when Lucille Cedercrans worked with the Master Rakoczi; and when Bruce Lyon worked with Djwhal Khul most recently. Here is the Master Djwhal Khul describing his perspective on transmission from remote distances during his work with Alice Bailey:

> The daily physical life of the Masters, of the Christ, and of those Members of the Hierarchy (initiates and accepted disciples) Who function in physical bodies, has had its orientation upon the subjective levels of life; the majority of Them, and particularly the senior Members of the Hierarchy, do not as a rule intermingle largely with the public or walk the streets of our great cities. They work as I do from my retreat in the Himalayas, and from there I have influenced and helped far more people than I could possibly have reached had I walked daily in the midst of the noise and chaos of human affairs. I lead a normal and, I believe, useful life as the senior executive in a large lamasery, but my main work has lain elsewhere— widespread in the world of men....
>
> This rule of solitariness or of withdrawing applies to all the Masters and to the Christ, for it is in the solitude of the mind, and as far as possible in the solitude of physical location, that the various branches of the great White Lodge [a term used for the Hierarchy by Djwhal Khul] have chosen to work ever since Atlantean days. It is not the solitude of a separative spirit, but the solitude that comes from the ability to be non-separative, and from the faculty of identification with the soul of all beings and of all forms. This can best be accomplished in the intense quiet of those "protected" areas where the Masters in the various branches of the Brotherhood have chosen to

---

98. This point should not be seen to conflict with the earlier discussion of masters incarnating through ashrams. Rather, it relates to the types of incarnational opportunity and freedom that are understood to characterise beings of these levels of advanced evolutionary development. Masters not incarnating through ashrams may choose to retain a physical body after liberation. Others may not. Those who incarnate through ashrams also may or may not retain the ability also to work through a physical human body, as is in keeping with their planetary service.

dwell. This solitude and physical isolation enables Them to work almost entirely from the level of the buddhic or intuitional plane, perfecting the Science of Impression [an advanced method understood to be used by the Masters of Hierarchy for the transmission of inspiration into humanity], influencing and working through those minds which are susceptible to Their mental impression. This applies equally to Masters in physical vehicles and to Those Who have 'no anchorage' in the three worlds [mental, emotional and physical-etheric bodies on the mental, emotional and physical-etheric planes]; it applies also to disciples who are in or out of the body, according to their destiny, immediate karma or form of service.[99]

And here is Djwhal Khul describing the actual transmission process during his work with Alice Bailey:

I wonder if any of you [the groups of students to which he was transmitting for a period] really grasp the extent of the effort which I have to make in order to reach your minds and teach you? When, for instance, I seek to send out these instructions I have to make the following preparation. First, I seek to ascertain the mental state and preparedness of the amanuensis, A.A.B. [Alice Bailey], and whether the press of the other work upon which she is engaged in connection with the Plan of the spiritual Hierarchy permits of her right reception; for if the work is exerting extreme pressure and if she is occupied with urgent problems, it may be needful for me to wait until such time as circumstances give her the needed leeway both of time and strength, and of mental detachment. My own sphere of occult work must also come under consideration. Then, having established a rapport with her, I have three things to do.

First, I must gather the group of disciples as a whole into my aura and so gauge its general condition of receptivity—for that must determine the scope of the intended communication. Do you realise, my brothers, that as you extend your power to grasp the needed lessons and learn to train your minds to think in ever wider and more abstract terms, you draw from me a correspondingly adequate instruction? The limitation to the imparted truth lies on your side and not on mine.

---

99. Bailey, 1957, p. 682-683.

Second, I must isolate in my own consciousness the extent of the instruction, detaching myself from all other concerns and formulating the needed material into a thoughtform which will be comprehensive, clear-cut, sequential in its relation to that which has already been imparted and which will lay the ground for the next instruction in due time.

Then third, I have to enter into that meditative condition, and that extraverted attitude which will enable me to pour out in a steady stream of constructive sentences which will express, to the mind of the amanuensis, the thoughtform as I see it and build it. Putting it otherwise, I become creative with deliberation and endeavour to convey to the vision, to the mind and to the intellectual perception of A.A.B. an ordered presentation of the thoughtform which embodies the lesson I desire the students to learn.[100]

And Djwhal Khul again more recently in relation to his transmission process with Bruce Lyon:

If one begins to redefine one's identity through 'identification' with the ashram or with Hierarchy itself, then there will of course be times when more or less of the Self is expressing through any one 'self', but there is no real separation of consciousness and identity. Only the concentration of that identity and its distribution through the component parts. We are one and yet together we are somewhat more than apart. This works in consciousness as well as form. So I see that it will help to explain further the mechanism by which such communications as this can take place—not only for your own comfort but for the satisfaction of our 'audience'.

The best way I can describe it is this—I 'shine' my consciousness towards and through your consciousness with intent. In this case the intent is the transmission of the third phase of the [Trans-Himalayan] teachings. Naturally this shining will stimulate your own consciousness and there will be a necessary period of 'preparing the field'. It is not dissimilar to the soul conditioning the three periodic vehicles [the mental, astral and physical-etheric bodies]. In this case it is my soul conditioning your soul if such a distinction can be made. This takes place upon

---

100. Bailey, 1955, p. 10-11.

the higher levels of the mental plane and so therefore requires that you are able to 'hold your mind steady in the light of the soul'. This then allows transmission to take place from my soul/mind to your soul/mind. Of course the transmission is energy and that energy will be conditioned by the quality and intent of my soul and mind as well as your own. That is why a clear alignment of intent is necessary. We need to be clear up front what it is that we are trying to achieve together. Then we need to have a functional mechanism that allows us to connect at specific times and in specific places (the [soul] body). Next we need to work on the quality of our connection—I am used to transmitting but must make adjustments to you as a receiver, and you must learn how to keep the channel clear and also how to purify your own fields so that they do not interrupt or distort the flow of information.[101]

Initiation, or the process by which evolutionarily more advanced souls periodically give sufficient empowerment to a younger soul so that there are signifigant shifts in their consciousness and capacity to hold higher rates of energetic vibration, is a ubiquitous law of the soul planes. Moving into deep collaborative relationship with a master, however, is not. This is an *opportunity* that is available to human beings on the soul planes, but not one that is necessarily engaged. To do so requires the inception of a multidimensional expression of the teacher-student relation that is greatly empowering, yet it is also deeply challenging. The presence of such inner masters is potent indeed. Having journeyed as far as they have, the intensity of their light is so great that it illuminates every space of one's own being, including anything that one has kept in the dark from perhaps even oneself. The expansiveness of their love is so full that the soul itself is humbled to near dust by the smallness of the love it is presently capable of demonstrating. The power of their presence is so awesome that in it, one may literally fear that one's own consciousness could be shattered. Yet simultaneously, choosing to do so and to witness and learn from the profound mastery they demonstrate in working with the kosmic and Earth energies, from full radical awakening and a titanic degree of evolutionary advancement, can substantially accelerate a soul's own path into supernova.

Contact between a master and a younger soul can be initiated by either side. This is important to note. It always occurs soul to soul, and an individual's capacity to bring through into waking awareness the content of such times of contact will be dependent on the extent to

---

101. Lyon, 2003a, p. 171-172.

which continuity of consciousness has been established between their soul and personality. The process allows teaching, empowerment, and guidance to be shared among beings operating on different planes and at different stages of unfoldment, so that the 'climbing selves' of younger beings are supported in their evolution and awakening by elders.

The experience of contact with liberated masters dwelling on the soul planes and on higher planes can occur in a 3rd person, 2nd person, or 1st person fashion. Sometimes we like to speak about this in terms of *energy, entity and identity*, respectively. The first of these, 3rd person contact, involves contacting the *energetic field* of a master. Like the corona of a sun, the energetic field of such a being is massive. In this form of contact one can access their energetic empowerment, or any needed information present in their field, without them needing to turn their conscious attention to the soul in question for a 2nd person relational exchange. Third person contact is therefore primarily energetic contact.

Second person contact involves, as we would expect, a *communicative relational exchange* between an entity, in this case a master, and a student. This is the kind of contact we might find easiest to understand, entailing the master offering teaching, guidance, and council to the younger soul. Yet it is one reserved for those whose alignment and identification with the purpose a particular master is serving, is sure, established, and without question. Masters are rays of Reality's Awakened Awareness that have penetrated into, chosen to hold a reservoir for, and to act as a transmitter of, kosmic energies and principles of Being that might contribute to a planetary scale empowerment of the realization of Earth's Purpose in the kosmos. As such, providing guidance to human beings is not the sole responsibility of such beings. Indeed, though all is interconnected and the evolution of one life-sphere on Earth has affects on all others, Hierarchy as a community has its own unique path of evolution just as humanity does. The masters, sages and bodhisattvas of Hierarchy are together pioneering into totally new spaces of collective kosmic evolution and awakening in consciousness, as well as the evocation of kosmic downloads that have the potential to contribute powerfully to the revelation of the sacred on Earth.

Lastly, 1st person contact involves a direct and immediate identification *with and as* the identity of a master via the heart chakra, which is the seat of the monad in the physical-etheric body. This is an electrifying, dynamic movement in which the master's identity arises and transmits forth through the very bodymind of the younger being so that the purpose, consciousness, and wisdom of the master are instantaneously clear. It is one in which there is a curious merging and yet simultaneous differentiation of identities between master and student. While the latter has moved sufficiently 'out of the way' for the master's presence to arise and operate through them, they retain conscious recognition and understanding of what is occurring. This is the phenomenon that Djwhal Khul described in his work with Alice Bailey as when some greater being *overshadows* a younger or more densely incarnated being. It also relates to a disciple's recognition that just as at an advanced stage of development and relationship with a master they are known esoterically as "a cheela [disciple] within the master's heart", so is it found increasingly at this stage that the master is also present in the younger soul's heart. It is by this method that the Trans-Himalayan teachings understand that Christ—the vastly evolved being referred to previously also by the name Maitreya—overshadowed and worked through the Master Jesus from the time of the baptism onwards. Hints of understanding this are found in the Adoptionist school of Christology in Christian theology.

One Hand Stretches Up — One Hand Stretches Down

As was noted previously, souls can also extend relationship into denser or subtler planes than the soul planes. When souls extend lines of relationship down into the lower planes (the mental, astral, etheric, and physical planes) they can transmit teaching to those they are in relationship with, which can then be distributed and shared in human culture and society. If that transmission extends further, and into the natural kingdoms (mineral, plant, animal), we have the stimulation and intuitive communion with the consciousness or sentiency inherent in these kingdoms—that of trees, plants, crystals, and stones.

When souls extend relationship into the monadic planes, as the great ashram of Hierarchy does continuously, they open a channel for communion with and the reception of transmission from Shamballa—a center of the planetary life we shall go on to explore in just a moment. As relationship is extended into the kosmos, it is understood that Hierarchy collectively holds open lines of connection with other greatly enlightened communities of lives on such planets as Venus, Jupiter, and Neptune, among others, as well as with the incredibly advanced lives who reside within the aura of the sun Sirius.

It therefore serves to make the point that as lines of relationship are opened up with communities of lives on multiple planes of reality, the soul can engage both a masculine or feminine orientation to that contact. This is the case in relation to access to the energies and information held on particular planes also. According to the masculine orientation, it can *penetrate* so as to reach out and make contact with various spheres of consciousness and energy or, according to the feminine orientation, it can *receive and embrace* so as to allow the identities, consciousnesses, and energies residing on those multiple planes of reality to transmit through it. Both of these should be understood as the two sides of a dynamic, magnetic, and radiant communion of life, consciousness, and energy, which is continually pervading the aura of the Earth. It is this communion that is transforming and transfiguring the planes that compose the Earth's aura, and radiating out into the kosmos our sacred narrative of becoming.

Relationship at the Monadic Altitude: Horizontal

Horizontal relationship at the monadic altitude becomes hard to speak of owing to a) the utter Self-identification of every monad with every other monad as the Self-perceiving energy of the Absolute (radical perspective), and of the kosmic Logoi within whose bodies they reside (evolutionary perspective); and b) that each monad on its own plane, deeper than persona and soul, operates unencumbered by the limitations of dualistic consciousness as we know it. Rather, monadic consciousness is characterized by a timeless, radically awake sentient awareness of everything, from mountains, plants and trees to each personality, soul and every 'other' monad, as the One Life of Perfect Godhead. And on the evolutionary side of the street, monads share the capacity for collective, instinctual response to, and transmission of, the Will, Love-Wisdom and Creative Drive of the evolutionary universe.

Shamballa

Just as we have spoken about the great ashram of awakened souls and masters in Hierarchy, who both act as stewards for all younger souls on the soul planes and guides for all humanity, on the monadic planes it is understood there also exists a community of deep planetary significance. This is a collective of beings who having stabilised their evolutionary center of gravity in the monadic bodymind, and who stand as far ahead of the great soul-elders of Hierarchy in their now kosmic realization and identification as the latter do to humanity. Together, such awesome presences stand as a radically awake black hole of power, transmitting our kosmic Purpose into the planetary sphere. Together, they form Shamballa.

From an Integral perspective, Shamballa can be understood as a LL collective culture and LR shared energetic field on the monadic plane, held by a community of radically awake buddhas whose UL and UR identification resides in the monadic bodymind. In terms of its precedent in the traditions, in the Tantric scriptures of Tibetan or Vajrayana Buddhism, Shamballa (spelt and known variously as Shamballa, Shambhala, Jambhala and Shangri-Lha) is discussed as a pure land existing in subtle energy above the Gobi Desert of Mongolia and China, where buddhas of supreme levels of planetary and kosmic realization dwell in their very subtle bodies. It is where Gautama Buddha is understood to have taught the Kalachakra Tantra, one of the unsurpassed Tantras that deals with the kosmic ecology within which we find our place, and the revelation of our primordial Buddha-nature throughout. In Hindu scriptures, Shamballa is the location in which it is understood that Kalki, the 10th Avatar of Vishnu, will be born. In the Tibetan Bon tradition, its founding Buddha, Tonpa Shenrab, who transmitted the sutra, tantra and Dzogchen teachings of Bon, is understood to have been born and come forth from Olmo Lung Ring, which is identified by the Bonpos as one with Shamballa. And for the Tibetan mahasiddha, Chogyam Trungpa Rinpoche, the Shamballa teachings inspired his vision of an enlightened society rooted in the dignity, warriorship, and wisdom of basic human goodness.

This reflects the teachings on Shamballa given out by those masters associated with the Trans-Himalayan teachings through such individuals as Helena Blavatsky, Alice Bailey, Helena Roerich, Lucille Cedercrans, and recently through Bruce Lyon. Here, Shamballa is understood as a center of the planetary life where buddhas and kosmic beings of profound enlightenment abide in an unbroken, dynamic identification with and transmission of the kosmic Purpose of the Earth, as well as the fire of divine Will to implement it. It is this Supreme Purpose that lies behind all manifestation, evolution and awakening on Earth and acts as the gravitational omega point outside of time towards which all is drawn, like suns and planets into a supermassive black hole. From this reservoir of dynamic, magnetic 'liquid Purpose', rivers of divine Will pour forth into Hierarchy, and increasingly also humanity. It is these that, in company with the forces of growth and unfoldment inherent in matter and consciousness, are understood to drive the evolutionary awakening of the whole planet.

Just as we saw above, in relation to our exploration of Hierarchy, where we spoke of masters incarnating through the ashrams that compose it, so also is this the case on a higher turn of the spiral with Shamballa. In the Trans-Himalayan teachings, at the core of the Shamballa planetary culture it is understood that there is a supreme Individuality whose immense altitude of evolutionary awakening and development is described as exceeding any other being within the aura of the Earth on the kosmic physical plane. This being is known in the Trans-Himalayan transmissions as Sanat Kumara.[102]

---

102. This is the name, meaning "eternal youth", of a being described in the Hindu

So great is the altitude of evolutionary awakening of Sanat Kumara that this being is understood to be incarnating through Shamballa as his monadic bodymind; the Christ and the planetary meta-sangha as his soul bodymind; and humanity and all the other kingdoms of the planet as his mental, emotional, etheric and physical bodyminds. He stands as the avataric representative on the kosmic physical plane of the planetary Logos of the Earth. This is the supremely enlightened being incarnating through the planet as a whole as its body of manifestation, and who also has such avataric representations on the kosmic astral and kosmic mental planes. It is via Sanat Kumara that the kosmic Will and Purpose of the Earth as it is envisioned in the heart-mind of the planetary Logos are transmitted into Shamballa, from that community into Hierarchy, into those pockets of humanity that are responsive to it, and into the body of the Earth. Djwhal Khul:

> From the standpoint of the forms of life in the four kingdoms of nature, Sanat Kumara is non-existent. In developed humanity, prior to moving on to the Probationary Path, He is sensed and dimly sought under the vague word "God." Later, as the life which the "seeds" have manifested reaches the higher layers or brackets in the human hierarchy, there emerges in the consciousness of the disciple, the assurance that behind the phenomenal world is a world of "saving Lives" of which he may eventually form a part; he begins to sense that behind these Lives there stand great Beings of power, wisdom and love Who, in Their turn, are under the supremacy of Sanat Kumara, the Eternal Youth, the Creator, the Lord of the World.[103]

Bruce Lyon on this same being:

> In Shamballa there are two groups, the Registrants of the Purpose and the Custodians of the Will. The first are responsible for alignment to the energy of Purpose emanating from the cosmic levels. The second are responsible for the implementation of that Purpose in

---

cosmology as one of the "mind-born sons of Brahma", the creator of the universe, and who is also described in the Theravada Buddhist *Janavasabha-sutta* as Brahma Sanatkumara, an eternally youthful being of vast realization who advises the gods of the higher planes to work with the teachings of the Buddha. He is also referred to in the Trans-Himalayan teachings under the name of Melchizedek, which means "King of Righteousness", and is the name of a being referred to in the book of Genesis in the Bible as "priest of the most High God" (Genesis, 14: 18-20).

103. Bailey, 1955, p. 288.

time and space through the release of Will energy from the reservoir of Will held in Shamballa. This reservoir is sustained by the commitment of Sanat Kumara to his purpose and supplemented by the will of the Solar Logos and so on. It is in effect the commitment of God to BE who he is in spite of seemingly difficult external circumstances. It is a commitment not to modify or qualify his essential nature....we might assume that it is this very quality of the Will-to-be that is needed to overcome the inherent resistances in the matter of this planet. The will does not modify and adapt the way that intelligence does. Nor does it love and educate the way that the soul or consciousness does. It rests in its own nature, the nature of essential divinity, and refuses to entertain any other reality. Sanat Kumara 'lives' in Shamballa on the monadic plane but he is not 'of Shamballa'. He is not 'of the cosmic physical plane'. His choice to Be here is freely given, and therefore the principle of Freedom lies at the very core of the planetary life.[104]

If we were to explore the nature of these beings who compose Shamballa according to both the radical and evolutionary vectors, we would recognize them as abiding in radical wakefulness to Absolute Presence; holding unbroken monadic identification as the dynamic Life and Self of the planetary Logos of the Earth; having free relationship with Sanat Kumara and other civilizations throughout the galaxy; as having penetrated into the kosmic astral and mental planes; and as operating through experientially kosmocentric altitudes of intelligence. Those who reside in Shamballa long ago came to full expression of their sun-ship as souls and consumed in fire even the last traces of karma at the soul level, which is really nothing else but the most subtle resistance to the Great Perfection of Awakened Awareness. They collapsed in on themselves in supernova, radiating and releasing their accumulated light into the atmosphere of their becoming so as to contribute to the transfiguration of the kosmic physical plane as a whole. In doing so, they left halos of wisdom-light that hold their transmission of awakening as available to all beings who come later who wish to contact it. They dissolved into black holes of monadic Life-force and then, just as the central cores of galaxies fuse when they come together, they surrendered the reservoirs of dark light that are their monadic bodyminds into one supermassive black hole. This is Shamballa, and it resides as a supermassive reservoir of awakened Life, kosmic Purpose and dynamic Power, at the center of which is the radically awakened heart-mind of

---

104. Lyon, 2003b, p. 193.

Sanat Kumara, the avataric representitive of the planetary Logos of the Earth on the kosmic physical plane.

To get some measure of just how potent the collective transmission of these beings is, it is understood that even the initiates and masters of Hierarchy cannot withstand at the current point in our planetary evolution anything other than temporary, cyclic exposure to the purity and raw voltage of their kosmic shakti. At present, this occurs at certain regular times of the year, though as the consciousness, bodies, cultures and societies of Earth and its beings increasingly evolve, greater exposure will become possible. Eventually, there may eventually come a time, likely some thousands of years or more in the future, when all life on the kosmic physical plane responds as instinctively to the pulse of Purpose and Will from Shamballa as do the limbs of the human body to nerve impulses emanating from the brain and central nervous system.

Relationship at the Monadic Altitude: Vertical

Vertical relationship at the monadic altitude involves those who compose Shamballa's feminine fertile receptivity and masculine, potent transmission. The receptivity side of their work involves sensitivity to the inflow of kosmic Will and Purpose from the vast heart-mind of the planetary Logos of the Earth on the kosmic mental plane, via Sanat Kumara, as well as other monadic Life-centered civilizations throughout the galaxy. In this respect, in the Trans-Himalayan teachings it is understood that those within Shamballa hold open lines of transmission with certain greatly awakened communities residing within the aura of the suns composing the constellations of the Great Bear, Draco, Orion, and also the galactic center. It is through the receptivity of those who collectively form Shamballa to these transmissions that evolution on Earth remains synchronized with that which is expressing in the larger field of our galaxy. It is through this held alignment that the dynamic peace, power, love and creative force of those intelligences are able to pour into our planetary process.

In terms of Shamballa's conveyance of energy into the rest of the planetary life, as previously stated, this involves the potent transmission of both radical and evolutionary awakening, of planetary Purpose, Will, and Life, which in company with the individual, behavioural, cultural and social influences of beings on the soul and personality planes drive the planetary evolutionary and awakening process. These streams of empowerment pour into all planes and kingdoms of nature (human, animal, plant and mineral) and deep into the subterranean fires of the

Earth at its core, which constitutes one expression of the planetary kundalini. This is something we will return to in a later section.

Earth is Eden

Chapter 12
Plane Access
Jon

This third sub-line of the evolutionary awakening vector relates to the unfolding capacity to penetrate shamanically into the various spheres of energy-matter that the planes represent, and to receive the energetic transmission of the communities of lives residing upon and within them openly. It also relates to the capacity to download the information they contain psychically, and to manipulate their substance as is the case in esoteric magic, or healing, for instance. Traditionally, these kinds of capacities have been the property of the shamanic traditions, such as those of Africa, Australasia and Siberia, as well as the native cultures of South and North America.[105]

As was said earlier in Chapter 10, on the subtler planes the radical Truth and evolutionary truths are more nakedly exposed and the differentiation between beings as apparently separated units progressively lessens. So, access to, or what we sometimes call *vision* into planes subtler than the physical naturally opens the door for access to the information found upon them, as well as the capacity to wield their energies and subtly journey upon them. In terms of the abilities that can be cultivated through plane access, this first begins with the capacity to see/sense energy, especially physical, etheric, astral, and mental energies. As it develops, one eventually gains the capacity to unfold subtle correspondences to each sense faculty (sight, hearing, touch, taste, and smell) so as to interact fully on all planes with increasing skill and ease. On the physical plane, one may begin to gain access to physical plane information in other locations (i.e. remote viewing) or

---

105. Clearly, the major religious traditions that stemmed from these early shamanic schools (i.e. Hindu Yoga) also carried certain capacities for this sort of plane access forward.

may be able to sense physical or emotional ailments of the physical body of another person simply by tuning into their field (i.e. medical intuitives). One also potentially gains the capacity for journeying on the etheric, astral, mental, and subtler planes out-of-the-body. This is something that most people experience in dreams but, as plane access on these levels is stabilized, one learns to journey on these planes whilst staying conscious and can consequently learn to acquire and initiate whatever information or contacts on these levels might be of greatest good.

DMT and the Eye of the Soul

One mode of opening up plane access (and relationship with beings residing on subtler planes than the physical) that is becoming increasingly prevalent in the West is through working with sacred plants, such as Ayahuasca, Peyote, Ibogaine, or certain forms of Psilocybin mushrooms. Many practitioners describe such plants as 'entheogens', which translates as "generating the divine within". These plants have the chemical, biological, and energetic capacities to produce powerful, long-lasting experiences of conscious access to the environments and energies of subtle planes, as well as relational contact with their inhabitants.

For most of us, access to and the experience of relationship on inner planes, such as the etheric, astral, mental, or soul planes, for instance, is not anything novel. The vast majority of human and animal beings experience access to and relationship on the etheric and astral planes during sleep, and every human being is at this very moment involved in various processes of light-work on the soul planes. As we said in the Chapter 10, it is the lack of integration between the various altitudes of identification—personality, soul, and monadic—that mitigates against a sufficient continuity of consciousness being experienced that our activities on each of these levels simultaneously are available to waking awareness.

So, what *is* novel, and what working with such plants offers the opportunity for, is to engage and train ourselves in such processes of subtle plane access and relationship *consciously*.[106] I would suggest that understanding one mechanism of action for this relates to an integration of ancient teachings concerning the pineal gland, and

---

106. This has relevance for development along the evolutionary lines of both multidimensional relationship cultivation, and plane access.

modern scientific findings on the naturally occurring psychedelic compound Dimethyltryptamine, or DMT.

The pineal gland is a small pea-sized structure in the bottom-center of the brain. For millennia, sages and philosophers such as Plato, Hippolytus, Yogi Bhajan, Swami Sivananda, and Descartes have suggested that the pineal gland has special significance as the point of interface for the soul in the physical body. In the words of Descartes, "In man, soul and body touch each other only at a single point, the pineal gland in the head".[107] Yet for all this time there has been a lack of rigorous scientific evidence to support anything close to such a position. In fact, until comparatively recently, it was a source of neuroscientific conjecture as to what the function of the pineal gland even was.

In recent decades, neurobiological and neuropsychological findings have allowed us to come to understand the pineal gland as an endocrine structure primarily important for the secretion of melatonin—a hormone that supports the regulation of waking and sleeping cycles in circadian rhythms—and to be composed largely of pinealocyte cells. Interestingly, pinealocytes, as they exist in the brains of non-mammalian vertebrates such as fish, amphibians, reptiles and birds, have a strong resemblance to the photoreceptor cells found in the retina of the eye. These are the light-sensitive cells that are responsible for phototransduction, which is the conversion of light into chemical signals the brain can process for visual perception. So, the first interesting finding to note here is the strong link in structural anatomy between the pineal gland and the eye.

More compelling research concerning the pineal gland began to emerge during the 1990s, and fascinatingly involved its possible relation to the endogenous production of DMT (endogenous means produced from *within* the body). Now, as noted above, DMT is a naturally occurring psychedelic compound, and its secretion produces hallucinatory experiences so potent that the individual in question often totally loses awareness of their physical surroundings and body in the intensity of the visionary journey. The major stream of academic knowledge concerning DMT and its *exogenous* psychoactive effects (exogenous means sourced from *outside* of the physical body) first came to light in the West as a result of Richard Evans Schultes' ethnobiological research during the 1950s, which documented its use by Amazonian shamans in their ritualized preparation, brewing and drinking of Ayahuasca.

Ayahuasca is one of the major plants of sacred use in South America and increasingly the world. It can be translated as the "vine of the soul" and it is composed primarily of a brewed combination of

---

107. Bailey, 1930, p. 43.

the *Psychotria Viridis* plant and the *Banisteriopis Caapi* vine. The former of these contains the major psychoactive component of Ayahuasca, which is DMT. The latter contains a number of chemical compounds known as Monoamine Oxidase Inhibitors (MAOIs). These latter prevent the DMT from being quickly broken down and metabolized by the stomach and lower intestine, thus allowing it to pass on unmetabolised through the blood-brain barrier so as to have the potent, long-lasting psychoactive effects it does.

In the last two decades of the twentieth century, more research began to emerge that took the collective intrigue concerning DMT to a higher level. Building on both the previous medical research of James Callaway, who theorized in the late 1980s that *endogenous* DMT production could be related to dreaming, and the teaching relating to the pineal gland transmitted by some of the world's spiritual traditions, in the 1990s Dr. Rick Strassman obtained ethics permission to conduct a five-year investigation into the psychological effects of DMT. Over the course of that study, which built upon the finding that the necessary constituents required to synthesize DMT *exist naturally in the pineal gland*, Strassman hypothesized that DMT is actually endogenously produced by the pineal gland. Furthermore, he theorized that a large number of visionary and mystical experiences reported by people over the course of history, such as those related to contact with supernatural beings, energies or entry into other realms in near-death experiences, for instance, could be physically explained in terms of occasions in which large amounts of DMT were produced by the pineal gland for unknown reasons, drastically altering the person's state of consciousness. Along these lines, the Taoist teacher, Mantak Chia, has suggested that the visions that arise in the practice of dark retreat, and which are understood to be so powerful in advanced stages of radical awakening practice, may arise through the endogenous production of DMT, brought on due to the isolation and sensory deprivation experienced.[108]

I would suggest that in light of the ageless wisdom teachings, these findings are definitely germane to our current discussion. This is because they provide an evidence-based scaffold upon which we can start to explore some of the premodern teachings that the pineal gland acts as the interface locale within the physical body for the consciousness of the soul. I would suggest also that a consideration of how evolutionary awakening unfolds through the various levels of our being over time may shed some light on why the evidence for this has been hard to come by and sporadic. Specifically, in the Trans-Himalayan teachings it is understood that it is not until a definitely advanced stage of development along the identification line that the soul bodymind is

---

108. Mantak Chia, n.d.

integrated and fused deeply enough with the physical body that we would expect there to be such major changes and activations within the endocrine system as are described here.[109] In his book with Alice Bailey, *Esoteric Healing*, Djwhal Khul gives a detailed teaching on the relation between the sets of seven major chakras in the etheric, astral and mental bodies, and the major glands of the physical endocrine system, with the correspondences as follows:

Chakra	Physical gland
Crown	Pineal gland
Brow	Pituitary gland
Throat	Thyroid gland
Heart	Thymus gland
Solar Plexus	Pancreas
Sacral	Gonads (testes for a male, ovaries for a female)
Base	Adrenal glands

Table 3. Table showing the glands associated with the seven major chakras.

Naturally, as deeper altitudes of UL soul and monadic awareness become embodied as radiant light, unconditional love and enlightened power in and through the physical body, we would expect there to be changes in the UR processes of hormone secretion within the endocrine system, and subsequent effects on behavior and experience. The changes explored above in relation to the pineal gland and its capacity for DMT secretion, which immediately opens up consciousness on the inner planes and the capacity for relationship on those planes, is one such example of this. While Strassman's hypothesis that spontaneously-occurring mystical experiences of inner plane access and relationship can be explained by sudden and unexpected releases of DMT from the pineal gland, I would suggest that a more stabilized form of this interaction is not just possible, but is the physiological basis of the third eye teaching held in so many of the world's esoteric traditions.

Teaching on a third eye, or a deeper capacity for sight available to human beings beyond our two physical eyes, which when 'opened' allows us vision into subtle planes of existence, beings residing on those planes, and some of the most profound mysteries of Existence itself, has a long lineage in the world's esoteric traditions. Within the Trans-Himalayan teachings, the third eye is understood as having two definitions—one fairly exoteric, and one more esoteric. The more

---

109. Likely the third initiation and beyond. We discuss the initiations in Chapter 17.

exoteric understanding of the third eye is that it relates to the brow chakra. The more esoteric understanding is that it represents a faculty of soul-seeing that comes online for a being when their personality—mental, emotional, and physical—is sufficiently integrated that it produces energetic changes in the brow chakra, which is physiologically expressed in the altered activity of the pituitary gland. The pituitary integrates and regulates the activity of the other sub-cranial endocrine glands throughout the rest of the body with their various relations to our mental, emotional and physical nature.

The second development understood to facilitate the opening of the third eye according to its more esoteric consideration is when the pure love-consciousness and wisdom-light of the soul have been sufficiently embodied in the etheric-physical, and specifically the crown chakra, that the pineal gland reaches a whole new level of activation. This, it is suggested here, brings online its stabilized secretion of DMT in a healthy and balanced way. When such a stage of development is reached that these two chakras and glands come into this intimate bioenergetic relation, it is understood that at *a third point within the head*—the third eye—opens. This allows not just conscious vision into inner planes as well as relational contact on those planes, but the capacity to actually see the soul, monadic and/or radical Absolute Reality-Light present at the heart of all forms on the physical plane in a manner that is a function of the seer's stage of development.

As development along this line proceeds, there can unfold the activation of inner correspondences to the senses of touch, smell, taste, and hearing, too. As subtle correspondences to all five senses are activated, one can imagine an individual unfolding not just the capacity for soul sight into and relationship with the light on all planes, but the ability to hear the sounds of all accessed planes simultaneously; to intuitively taste and smell the flavor and odor of all consciousness and energy-matter of every accessed plane; and to be able to touch, physically, energetically, with consciousness and the empowering current of monadic Life-force, the consciousness and forms of all accessed planes.

The above involves the unfoldment of more subtle correspondences to those inner capacities for contact that are inherent in the personality vehicles also—mental, emotional, and etheric—such as telepathy, clairvoyance, clairaudience, and psychometry, for instance. As has been the testament of yogic and shamanic traditions for millennia, all of these capabilities are trainable in human beings.

Access to the LR Energy Fields of the Planetary Cultures

Each of the major LL planetary cultures that we considered in the previous chapter has its corresponding LR energy or physical field. In relation to the mineral kingdom, this is the geosphere. In relation to those spheres of evolutionary complexity composing all biological life (planet, animal, human), we speak of the biosphere. In relation to the field of all human thought on the mental plane, Pierre Tielhard de Chardin coined the term "the noosphere".

As plane access is unfolded onto the soul planes, one comes into contact with the energetic field of Hierarchy—the planetary meta-sangha. Here, one gains access to soul-altitude environments in which information relevant to much larger spheres of time and space within our evolutionary history becomes available. In relation to the temporal aspect of this, there now comes online the capacity to see back into the soul-altitude history of oneself in terms of one's previous incarnational history, and that of others, as well as to clearly see the threads of karma and destiny that may determine that which occurs in the future. In terms of the spatial aspect, the capacity for higher mental, buddhic, and lower atmic plane journeying both within the aura of the Earth and out into the solar system come online; it is in this context that we may hear of genuine examples of subtle travel to other planets and further. At this altitude of plane access, there also develops a much deeper capacity for healing work as the magnetic, pure love energy that is the basis of all being within Hierarchy on the soul planes becomes available for reception and transmission.

As plane access continues to unfold, and a being's capacity to contact and then wield the energies of subtler planes is extended into the monadic, one comes into contact with the LR energetic field of Shamballa, and as a result gains access to the transmission of planetary Will, Purpose, and Life. As this contact is stabilized, and one's identification line also shifts into the monadic altitude, we speculate that there emerges the capacity to magically manipulate the structural composition of all seven sub-planes of the kosmic physical plane, including physical reality also. As of yet this remains a theoretical hypothesis, but its basis is that it is on the monadic altitudes that Absolute Reality expresses with greatest power, and it is this depth of power that is required for the manipulation and transformation of the densest level of matter.

An important point to take from all of this is that access to the energy-matter and information of a particular plane does not necessarily require or evoke a shift in identification to the same altitude. It does open the door to it, *but it doesn't automatically take you through it.* This means that it is quite possible for individuals and cultures to gain access to the information extant on subtle planes without necessarily having unfolded the wisdom, love, and intelligence that are the property of their corresponding altitudes of relative identification and intelligence. This point opens up important doorways in terms of the way we understand the difference between evolutionary, ancient, and ageless wisdom, especially when considered within the context of all the lines of the evolutionary awakening vector, and the vector of radical awakening also. This is something we go on to explore in a later chapter.

Hylozoic Holarchy and the Planetary Chakras

According to Ken Wilber's expansion on Arthur Koestler's notion of the holon, reality can be considered to be composed of neither fractured parts nor of one whole, but of a hierarchy (or, to use his terminology, 'a holarchy') of holons. These holons are at once parts and wholes simultaneously—wholes that are parts of greater wholes, such as an atom within a molecule, a molecule within a cell, a cell within an organism etc. Within the Trans-Himalayan teaching this same perspective plays a central role, but from a kosmic hylozoic perspective.

Hylozoism is the name of a worldview, prominent in the ancient world, in which the entire universe is understood as *alive*. From the hylozoic perspective, there is therefore no such thing as inanimate matter anywhere. When this vision is plugged into the holarchical understanding of the kosmos put forward by the Trans-Himalayan teachings, each and every holon is understood as a living, sentient being that is both a whole in itself and a center of force within the body of a greater, more inclusive being. Indeed, the Trans-Himalayan cosmology sees the whole universe as composed of an infinite hierarchy of beings that are wholes in themselves, and yet chakras within greater beings who are in turn wholes in themselves, and yet chakras within greater beings—all within the Boundless Awake Matrix of Primordial Presence.

Using this teaching as a jump off base, and situating our consideration within the sphere of the planetary holon that is the Earth, the three major centers of LL relationship described in the previous chapter—Humanity, Hierarchy, and Shamballa—embody not just LL planetary cultures that have their own LR energetic fields on the kosmic physical plane *but UR chakras within the etheric body of the planetary Logos of the Earth.* This highlights the point that the constituent elements of the UR quadrant—from atoms to molecules to cells to organisms and then organs such as a reptilian brain stem or limbic system—are themselves always holonically composed of smaller elements that are wholes and yet parts of greater wholes. So also is that the case when we are looking at AQAL matrices for beings incarnating through ashrams, planets, solar systems, etc. That is, from the hylozoic Trans-Himalayan perspective, their bodily structures are composed of communities of lives, whether that be in relation to the lives composing humanity, which are understood to embody the throat chakra of the planetary Logos of the Earth, the cells of an animal's limbic system, or the atoms whose bonds define a molecule.

Humanity—Earth Shaman

In the Trans-Himalayan teachings, and in relation to a further consideration of the contents it offers to the LR quadrant of an AQAL matrix, humanity occupies a unique position within the various kingdoms, or spheres of evolutionary complexity, which compose the Earth. Those kingdoms are described as seven. The three pre-human kingdoms are the mineral, plant, and animal kingdoms. Humanity is the fourth of the seven, making it the middle kingdom. The fifth kingdom is composed of the awakened souls of Hierarchy that presently resides on the soul planes (higher mental, buddhic, and lower atmic planes). The sixth kingdom is formed by a community of lives that are the most advanced bodhisattvas of Hierarchy, whose role it is to maintain a bridge in consciousness and energy between Hierarchy and Shamballa through the lower and higher atmic planes. The seventh kingdom is the fully awakened monads of Shamballa that presently reside on the monadic planes (higher atmic, monadic, and logoic).

The uniqueness of humanity's position within this scheme relates to its potential to serve as a bridge between said pre- and post-human spheres of evolutionary complexity. To the degree that this opportunity is engaged, this may allow humanity as a whole, as well as the animal, plant, and mineral kingdoms to be increasingly transformed and transfigured by the radiant love-wisdom and dynamic empowerment that flows from the post-human kingdoms. It might allow the UR and LR forms of these kingdoms to be stimulated sufficiently in their evolution that they are increasingly able to house the higher altitudes of UL and LL consciousness and monadic Life that characterize the post-human kingdoms. Profoundly, this would allow eventually the emergence of the post-human kingdoms as *biologically* embodied life-spheres on Earth.

In order to serve this process, humanity is called to serve as a planetary shaman, penetrating into the planes upon which the post-human kingdoms of life reside and acting as a vehicle for their transmission into the natural kingdoms. It is this that may eventually allow humanity to stand as "the love that can hold all kingdoms in right relationship with each other".[110] According to the Trans-Himalayan teachings, it is this achievement—the irrigation of all multidimensional inter-kingdom relationships with the waters of pure love—that will open the way for the unfoldment of the *kosmic* Purpose of humanity and the Earth. What is this Purpose? Keep reading...

---

110. Lyon, 2010, p. 21

Chapter 13
Lines of Intelligence
Dustin

As mentioned in Chapter 1, Wilber often lists three basic groups of lines. They are self-related lines, cognitively-based lines, and talents/skills. Thus far, in the previous three chapters, we have examined three self-based lines (identification, relationship, plane access) according to how they are understood to move through various planes of reality in the Trans-Himalayan teachings. Here, we wrap up the discussion on evolutionary lines with a consideration of how these notions are informed by conventional Integral Theory. In principle, as we have tried to do thus far, this linking with contemporary Integral Theory will further show how the work we have outlined so far might be supported. The previous three chapters traced lines of development as they progressed from the physical plane through the emotional plane, through the mental plane and onwards, into buddhic, atmic, monadic planes and beyond.

In Integral Theory, the lines of development studied have particular reference to the intelligences of the personality (physical, emotional, and mental). As much of the research available falls into the category Wilber calls 'cognitive-based lines', it inevitably tends to relate to the mental plane. When lines of development unfold on the mental plane, they express in a much more lens-like fashion; they express in the form of *worldviews* through which we understand and interpret the world, such as is encapsulated in Spiral Dynamics. As we see below, the configurations of various lines of development serve to literally create the world that we see, experience and interact with. Because most of humanity resides within the most common mental structures, it is important that we explain here the various stages of consciousness that can unfold on the mental plane and how these stages affect our everyday experience of reality. These varying worldviews, as expressed in this chapter in relation to the mental plane, likely show up quite differently on kosmic paths where beings are operating through structure-stages of development so advanced that their relationship to worldviews may no longer be as relevant.

Conventional Lines of Development in Integral Theory

As he outlines in the book *Integral Psychology*, Wilber uses a theory of multiple intelligences as a foundation to design a tool called the Integral Psychograph.[111] The Integral Psychograph offers the reader a simplistic pictorial representation of multiple areas of development.

The graph below elaborates further on the notion of the Integral Psychograph. It depicts six major conventional lines of psychological development (cognitive, values, orders of consciousness, self-identity, worldview, faith). Although there are indeed some exceptions, notice how the majority of the lines of development fall within the range of the personality; most stages fall within the more surface layers of the self as they express upon the physical, emotional, and mental planes.

Figure 20. (on next page) Chart showing different lines of intelligence unfolding though the Integral structures and planes. (Adapted from Wilber's *Integral Spirituality*[112])

---

111. Wilber, 2000b.

112. Wilber, 2000b.

Altitude/Plane of Evolutionary Awakening

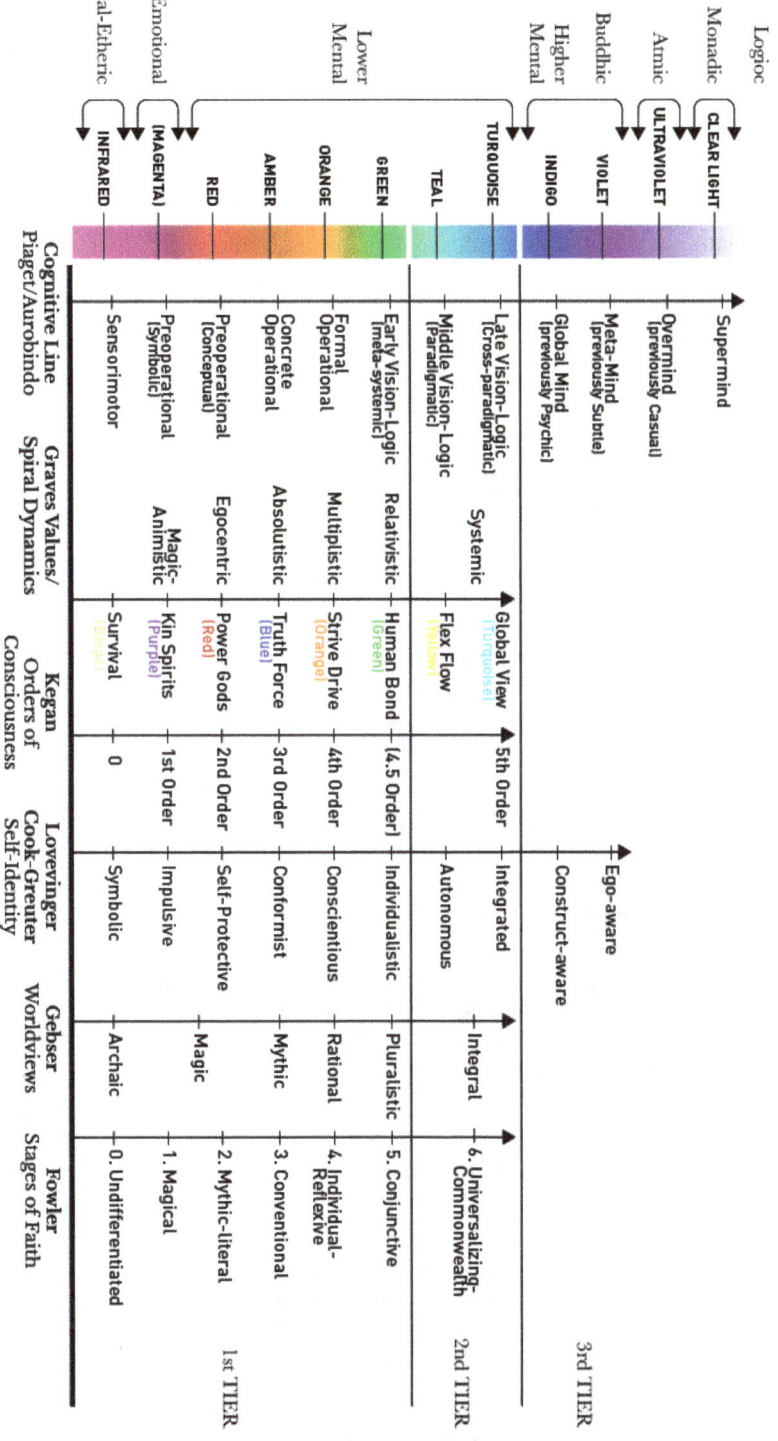

Planes: Physical-Etheric · Emotional · Lower Mental · Higher Mental · Buddhic · Atmic · Monadic · Logoic

Altitude colors: INFRARED · (MAGENTA) · RED · AMBER · ORANGE · GREEN · TEAL · TURQUOISE · INDIGO · VIOLET · ULTRAVIOLET · CLEAR LIGHT

Conventional Lines of Development

Cognitive Line Piaget/Aurobindo	Graves Values/ Spiral Dynamics	Kegan Orders of Consciousness	Loevinger Cook-Greuter Self-Identity	Gebser Worldviews	Fowler Stages of Faith
Sensorimotor		0	Symbolic	Archaic	0. Undifferentiated
Preoperational (Symbolic)	Magic-Animistic	1st Order	Impulsive	Magic	1. Magical
Preoperational (Conceptual)	Egocentric	2nd Order	Self-Protective	Mythic	2. Mythic-literal
Concrete Operational	Absolutistic	3rd Order	Conformist	Rational	3. Conventional
Formal Operational	Multiplistic	4th Order	Conscientious		4. Individual-Reflexive
Early Vision-Logic [meta-systemic]	Relativistic	(4.5 Order)	Individualistic	Pluralistic	5. Conjunctive
Middle Vision-Logic [Paradigmatic]	Systemic	5th Order	Autonomous	Integral	6. Universalizing-Commonwealth
Late Vision-Logic [Cross-paradigmatic]	Global View		Integrated		
Global Mind (previously Psychic)			Construct-aware		
Meta-Mind (previously Subtle)			Ego-aware		
Overmind (previously Casual)					
Supermind					

Graves Values / Spiral Dynamics colors:
Survival [Beige] · Kin Spirits [Purple] · Power Gods [Red] · Truth Force [Blue] · Strive Drive [Orange] · Human Bond [Green] · Flex Flow [Yellow] · Global View [Turquoise]

Tiers: 1st TIER · 2nd TIER · 3rd TIER

Just as we have seen with other aspects of the evolutionary vector, stretching along the y-axis of the graph a colored spectrum ranges from infrared to clear light, indicating altitude. Next to each colored altitude you can also see listed the various planes of existence (physical-etheric, emotional, mental, buddhic, atmic, monadic, and logoic) as articulated in the Trans-Himalayan teachings. In other words, the graph shows that the higher a particular skill reaches in altitude the more it will express faculties of intelligence present in deeper structures of the relative self-system (personality to soul to monadic). Notice also that the majority of stages for the major lines listed above fall within the mental plane (red-turquoise).

Additionally, just as we saw with the more esoteric aspects of the self-related lines in both the evolutionary and radical awakening vector, each of the conventional lines listed above develops relatively independently of all the other lines.[113] That is to say some lines might be well developed, while others are underdeveloped. For example, explaining one's own Integral Psychograph within the mental plane, one might say "I have high cognitive capacity but sometimes feel as if I have a low level of social intelligence."

The Integral Psychograph imparts several benefits to any type of analysis. First, it allows easy reference when comparing stages of development within a single line. For example, we could say that in any particular line of development that "green pluralistic" altitude in the mental plane is higher than "orange rational" altitude in the mental plane (each of these terms will be explained in more detail shortly.) That is, *pluralism* transcends the limitations of the *rational* level of development yet includes all of its important contributions, whereas the reverse is not the case (orange rational does not include the perspectives of green pluralistic thinking). In turn, orange represents a higher level of complexity than amber or red, and so on.

To give a second example, the psychograph is also useful if we wish to investigate a single altitude across various lines. Examining the graph at one particular altitude (or color) provides us with a general orientation of where a stage in one line rests in comparison to a stage in another. Although each researcher uses different criteria to determine development in each line, we can see by looking at the graph that Loevinger's "conformist", Fowler's "conventional", and Piaget's "concrete-operational" all are contained within the amber altitude on

---

113. Other theories (e.g. Kurt Fischer's *Skill Theory*) claim that development might be more like a "web" of particular skill sets rather than clean and separate areas of linear intelligence. This book agrees with this notion to a certain extent in that our articulation of religious orientation is indeed itself a particular "web" (or as we call it a bundle) of various intelligences.

the mental plane. So although we cannot say that all three of these levels in the various lines are the same, we can indeed begin to intuit a similar flavor within the various stages. Both of these benefits (i.e. the ability to compare stages within a single line and the ability to investigate a single altitude across various lines) show us how a graphic representation of complex matters can allow us to make reference quick and easy.

The Rainbow of Enactment

When understanding the significant role these mental plane lines of intelligence play in our lives, it is important to note that the specific combination of development across multiple lines results in a particular lens through which an individual views reality. As stated, this is particularly true for human beings who tend to have a center of gravity in the mental plane and identification as a personality. Different levels of development, across different lines, enact totally different world-spaces. In order to discuss the point in more general terms we can simplify the developmental spectrum and focus on four stages of development within the mental plane to see how each brings forth reality in a different way. Here we look at the traditional altitude, the modern altitude, the postmodern altitude, and the integral altitude. When dealing with a deeper level of granularity within the mental plane we sometimes use the term structure, or structural development, to denote a specific stage of evolutionary unfolding across any of the four types' evolutionary lines.

Following Wilber's lead, each stage corresponds to a particular color or altitude related to the personality and shows up as a worldview through which the mind interprets the world. The traditional level of structural development corresponds to the color amber on the rainbow spectrum. The modern level corresponds to an orange altitude. The postmodern level corresponds to a green altitude. The integral level corresponds to a turquoise altitude. Each of these levels gives rise to a particular worldview. Each of these worldviews has both healthy and unhealthy expressions around the world today. For those interested in participating in collective evolution, the work is to facilitate the adaptation of their pathological expressions into healthy expressions. Just as when an individual is functioning at optimal potential all levels of their being are in sync and healthy, so too the human species functions at optimal potential when each developmental view is as healthy as possible and is coordinated into a larger matrix.

Let us now turn for a moment to a more detailed look at each worldview as it arises from the various levels of structural development from traditional, to modern, to postmodern, to integral.

A person with a traditional level of structural development (roughly 65% of the world's population, according to Wilber) enacts a traditional reality. Often an individual at a traditional level of structural development takes a dogmatic stance to his or her own belief system. Often he or she focuses on an absolute "truth" and employs a puritanical sense of right and wrong. Usually, an individual at this stage of development is willing to control impulses in exchange for deferred fulfillment (being 'good' now is perceived to lead to rewards later). The individual with a traditional level of structural development often shows conformist behaviors and indicates that the approval of his or her group is of the utmost importance. He or she is often kept in order through feelings of guilt. In many cases, the individual's own identity extends to that of his or her own group, family, political, social or religious community, while seeing the views of those outside the group as either wrong or out of line with the one *true* path. When enacting a traditional reality, the modern world, the postmodern world, and the integral world do not exist. We might say that the traditional person is wearing amber-colored glasses and can only see an amber world.

In a similar way, the modern person wears orange glasses and enacts only an orange world (roughly 20% of the world's population). Individuals at a modern level of structural development begin to question and examine all of their existing beliefs. They begin to scrutinize the myths they believed without hesitation at the previous stage, in order to find deeper meaning. For the first time, individuals recognize the ability to have their own opinions outside the restrictions allowed by the group or scripture.

Due to the pragmatic and reflective nature of this stage, individuals may become agnostic or atheist, both of which represent healthy expressions of structural development at this stage. The individual at this stage enacts a reality in which a strong emphasis on autonomy, independence and success is of the utmost importance. He or she is usually emphatic about embracing the value of the scientific method, evidence, and *tried-and-true experience*. At this stage, individuals gain a much fuller capacity for 3rd person objectification and reflection upon their own thoughts and beliefs. This means that one moves beyond blind belief, in particular religious/secular ideologies, to *operate* on them to improve them consciously and critically. Often folks at this stage of structural development have little access to penetration into the soul and monadic planes (either because penetration was never developed or, in some cases, penetration developed at an earlier stage but the physical plane/Gross state bias of this modern structure now prevents such access). For the modern individual, the postmodern and integral worlds are over their head. They lack the developmental capacity to bring forth the level of sophistication and perspective-taking necessary to enact a green world at the next stage.

A postmodern person wears green glasses and enacts a green world (roughly 10% of the world's population). The individual at this stage begins to realize that life's issues do not have to be black and white. The individual becomes comfortable with, and may even enjoy, the embrace of paradox. Individuals with a postmodern level of structural development on the mental plane might claim that all traditions are simply different perspectives of the one Ultimate Reality. The religious pluralism expressed from this stage moves beyond mere tolerance, as expressed in the modern stage, to actually fully embrace other religious traditions for their inherent value. Individuals at this postmodern level of structural development begin to recognize the cultural embeddedness of their own religious and spiritual beliefs. As a result, they often begin to search out other spiritual systems. They search not with a desire to convert those of other faiths, however (as might be the case at a traditional level of spiritual intelligence), but in order to take in other perspectives, and to find out how another's view may be able to supplement their own. When a green world is enacted, it is full of a richness of diversity and multiple perspectives, but often loses a sense of organization and holarchy that can lead to effective and dynamic transformation of the systems involved.

An individual with an integral level of structural development wears turquoise glasses and enacts a turquoise world (less than 5% of the world's population, according to most estimations). Having taken the perspectives of other religious traditions, supplemented their own beliefs, and uprooted their own worldview from the limiting perspectives of his or her own culture (to whatever degree possible), the individual at an integral level of structural development begins to find a sophisticated yet simple orientation. Our colleague Clint Fuhs calls this '2nd Simplicity', thereby distinguishing this simplicity that rests on the other side of postmodern complexity from the first type of simplicity that arises at more traditional levels of growth.

Those at an integral stage have found a stable center within themselves with regard to their own personal beliefs. Often this level of structural development begins to include some degree of vantage point development in the radical awakening vector (at higher integral stages of intelligence unfoldment, beyond the mental plane, at least some vantage point development through the vector of radical awakening is required).

Most fundamentally, the integral stage of structural development recognizes the importance and value of all preceding levels. This is an incredibly shift, as before this stage of unfolding *each previous stage exists in conflict with the perspectives of all other stages as to who is singly 'right'.*

For example, an integral level sees that the postmodern altitude served as a filter to neutralize all dominating tendencies. Passing through the postmodern level ensures oppressive tendencies do not resurface when healthy, natural hierarchy returns at the integral level. Ultimately in its fullest expression, an integral level of structural development would not only try to include all four vectors of spiritual intelligence outlined in this book, but would also be able to make each of the distinctions operational for the benefit of all.

It is important to note that the evolutionary vector of development does not stop here, but rather continues into other integral stages. In relation to the unfoldment of this developmental journey, all human beings begin at square one and all human beings stop growth at various stations along the spectrum of vertical development. Growth continues or stagnates along the path for various reasons. In some situations, growth comes to a halt because culture, society or a particular belief system limits the human potential of an individual. In other circumstances, challenging life conditions may cause a person to focus on survival needs rather than aspects of being that might promote their own flourishing. In situations where growth does continue, it may be a result of mentorship or coming into contact with a teacher who is more developed. In other situations, growth might continue as a direct result of a tradition or belief system mapping out the full spectrum of human possibility. We hold the intention that this book falls into the category of a successful "growth catalyst" as well.

As a result of mixed levels of development, today we are left with a multitude of human beings, all wearing different colored glasses and all of whom are enacting different colored worlds. If we are to make any sort of attempt at articulating a global vision of unification, it is vital that we understand the way that varying developmental levels of intelligence unfoldment (along with factors in the Four Quadrants) bring forth different views. Otherwise, we would be lost when trying to explain a specific reality to an individual(s) who culturally, socially, or developmentally is wearing glasses that prevent him or her from enacting it. A person at a traditional level of development reading this book, for example, may not be able to see the same world as it was enacted from the perspective it was written. If they are to have any positive and lasting effect outside of their own memetic/developmental construction of reality, the ideas in this book would need to be translated into a language that the person can hear and receive. This capacity to translate comes online even more fully at the integral stages of development.

Further Reaches of Conventional Lines in Integral Theory

Some research shows that there are interesting phenomena that unfold as certain lines reach higher in altitude. For instance, in Chapter 15 we trace Wilber's self-reporting of the higher stages of cognitive development that extend into and are the intelligences of the soul and monadic altitudes of identification. There we find a bridge between some of the more esoteric aspects of evolutionary awakening and the progressive aspects of Integral Theory coming together. In his descriptions of structure-stages, Wilber refers to vision logic, illumined mind, intuitive mind, overmind and supermind. Here, when considering these structure-stages in light of the Trans-Himalayan teachings, we see the cognitive line of development extending beyond the mental plane and into the buddhic, atmic and monadic planes of evolutionary awakening.

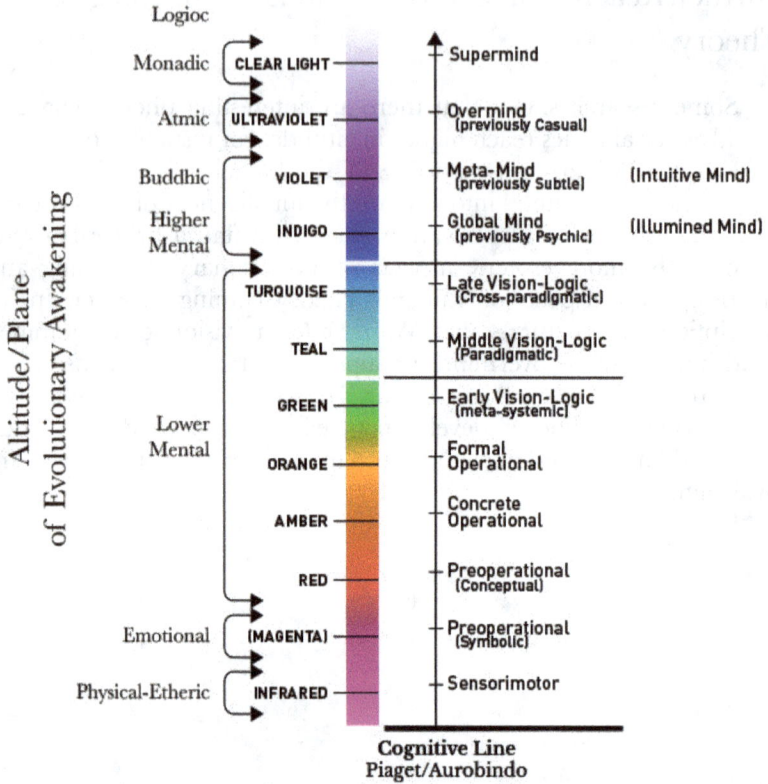

Figure 21. Chart showing correspondences along the cognitive line of intelligence development between the Trans-Himalayan teaching on the planes and conventional Integral Theory

Although it is speculation, we suspect that certain capacities come online as particular lines of development reach into the higher mental, buddhic, atmic, monadic, and logoic planes, hence constituting the third category of lines that Wilber calls talents/skills. For instance, certain forms of telepathy and other capacities for direct perception and intuition may be an expression of particular conventional talents/skills extended into higher planes (e.g., buddhic, atmic, monadic, logoic) or alternatively these lines extended into the highest available capacities of particular altitude of identification on a particular plane. In this way, we might venture to guess that there are actual siddhis (spiritual capacities) that can develop in nearly every line if it is allowed to mature to its full evolutionary unfolding.

Plane Specific Lines of Development

The second point worth noting is that particular lines of intelligence might be best understood as related to particular planes. This ties in with the Trans-Himalayan perspective on how the different intelligences develop. There, it is understood that the various lines of intelligence unfold as the UL consciousness of a particular altitude of identification (personality, soul or monad) is focused in a particular UR body (physical-etheric, astral, mental, the soul body, etc.) and working with the LR fields of energy-matter of a particular plane. Naturally, of course, this all occurs within a shared LL culture (Humanity, Hierarchy, or Shamballa). For instance, lines of development like empathic intelligence and kinesthetic intelligence are best associated with the emotional plane and physical plane respectively. These lines of development are not therefore necessarily mental capacities associated with the mental plane, as are so many of the other conventional lines of development. In this way, it may be the case that as an emotional line continues to develop it does so through various stages within the emotional plane itself (without necessarily moving in to the mental plane). The same is true for physical or kinesthetic intelligences. It may be the case that the physical plane itself has a whole host of lines of intelligence with their own respective developmental stages within it. This means that further development on the physical plane may or may not move vertically into the emotional plane, and so on.

PART 5: VECTORS AND LINES: FURTHER EXPLORATION

Within the robes of the ever becoming the always already is remembered. Once the experience of radical awakening takes place in some stable way then the journey of becoming continues, but becomes a blissful creative dance.

(Master Djwhal Khul and Bruce Lyon)[114]

---

114. Lyon, 2010, p. 54.

Earth is Eden

Chapter 14
Evolutionary, Ancient, and Ageless Wisdom
Dustin and Jon

In this book, we have offered a two-vector model for understanding awakening and evolution that extends both the Trans-Himalayan teachings and conventional Integral Theory. In this chapter we offer a meta-framework for how this model allows us to understand wisdom, knowledge and insight as they have been generated by humanity through time. By way of us offering a disclaimer, we understand that the criteria we are using to explore this meta-framework arise from the confluence of our own relative perspectives with those inherent in where humanity stands at its current point of evolution. As such, of course this meta-framework is as much an expression of our own orientations as well as what we believe to be a vital piece of the coming universal spirituality.

Wisdom teachings from the great traditions are often sourced in antiquity. When we work with such information, we must employ a system by which we can preserve the relevant insights of the traditions whilst simultaneously jettisoning those elements that are either culturally irrelevant today or simply based in pre-rational structures of mental plane intelligence that are no longer serving. In order to set up the hermeneutic of insight that we employ here, it is important for us to offer a distinction between 'evolutionary', 'ancient', and 'ageless' wisdom. This is an approach that the model we use here can well accommodate, and although these distinctions are simplified generalizations, pointing to them has heuristic value.

One approach to the validity of a perspective and its inherent 'wisdom' sees that validity as existing less in humanity's historical past and more in the future. This is the 'evolutionary' perspective most common in modern Western culture. According to this perspective, the accuracy of our collective wisdom and our capacity to assess reality has increased over the course of evolution. With more complexity and a greater number of developmental perspectives, the argument states, we move more and more into our human potential. Further, owing to the fact that higher developmental views often transcend and include the perspectives of lower views, the position is often held that more evolved views hold more reality than those views that are less evolved. This is a valid point as far as it goes, and one that Integral theorists are sure to take into account. One of the common criticisms leveled at this perspective, however, is that its linear evolutionary orientation fails to recognize and honor previous periods and cultures of history in which 'wisdom' of an extraordinary nature existed. This sort of wisdom remains rare today, and in some cases is still unknown to the masses regardless of the average center of gravity in terms of evolutionary intelligence.

A second orientation, often held in opposition to the first, sees greater wisdom existing in the past when compared to today's levels of spiritual competency. This is the 'ancient wisdom' perspective. This perspective looks back to ancient, pre-modern times and places such as Ancient Egypt, South America, India, and Tibet, for instance, and sees the flourishing of 'golden ages', uncontaminated by the trappings of modern and post-modern society (and the rational/egoic reductionism that often accompanies them). This view recognizes that certain levels of access and vision into the inner planes were fully integrated into societies of the past. It also recognizes that many cultures of the past had systemized and sacred rituals that helped a self to locate itself in the kosmos and ascertain where on the developmental journey it might be. A criticism often leveled at those who take this stance is that they are overly-romanticizing past periods of history when in fact the general altitude of consciousness development was ethno-centric, pre-rational, and heavily invested in magical and mythical thinking.

A third orientation acknowledges that some insights are based neither more in the past nor in the future, but rather are eternal. This orientation takes an 'ageless' wisdom perspective. From this point of view, individuals acknowledge that regardless of time period, culture, or any other temporary and fleeting construction of mind, relative reality can always be deconstructed to its Absolute Base (radical awakening). An unbounded Nondual Wholeness is available, waiting to be recognized by any human being willing to explore the true nature of his or her own awareness. Although this view is less contested, criticism is sometimes made that it has an Eastern bias that gives preferences to the insights of Hinduism and Buddhism.

The perspective we take in this book honors the truth in *all three* approaches, depending on the content and context of what it is being used to explore. The 2-vector methodology employed herein recognizes that from a Western point of view, an evolutionary orientation (i.e., that our view of the world has grown more accurate as development increases) is true in many cases when addressing the issue from the perspective of the evolutionary awakening sub-line of intelligence unfoldment.[115] From this perspective, in terms of the degree of the sophistication inherent in the lens of intelligence through which an individual or group is enacting the world, criticism of the ancient wisdom perspective has definite validity. It is *not* to the dignity of humanity as it stands today and the struggle to grow and unfold higher, more inclusive perspectives we have passed through, to champion pre-rational, ego- to ethno-centric perspectives. Development of higher, more inclusive views is real.

However, from the perspectives of radical awakening, identification, relationship cultivation, and plane access, much of the Occidental world suffers from stagnation at best and pathological regression at worst. Although advancements into modern and postmodern structures of mental plane intelligence are positive, they have come with a massive level of physical plane reductionism. Deeper altitudes of identification beyond the personality bodymind (the soul and monadic levels) are not acknowledged; nor are capacities to cultivate relationship with beings on, and to penetrate into, planes more subtle than the physical. In this sense, there may be much to learn from the perspectives held by cultures that remain less developed according to a mental plane intelligence structure-scale. Although their development along the ladder of mental plane intelligence complexity may be less, insights into ageless and ancient wisdom may still far transcend those present in contemporary Western culture.

In relation to ageless wisdom, we need to look no further than the Tibetans as an example here. Although most individuals in the Tibetan culture still operate at magical and mythic levels of structural development, they remain the wisdom keepers of some of the most sophisticated teachings on radical awakening available today. Or in relation to realm access, we could give the example of the Kogi tribe of Colombia. The shaman priests of the Kogi, or *mamas* as they are called, are raised and taught from birth in caves where there is virtually

---

115. This is particularly the case in relation to the mental plane. There are numerous examples of premodern cultures who display forms of intelligence unfoldment related to the physical plane, for instance, that are definitely more advanced than those conventionally known in modern and post-modern Western society. The Hatha Yogis of India are one example here, as are the Australian Aborigines, who have demon-strated intelligences related to navigating wilderness areas of land far in advance of Westerners.

no light or sounds for the first nine years of their life so that they can gain expertise in accessing subtle planes before they even have any real experience on the physical plane. Just as the Tibetans and the Colombian Kogi, among others, have helped to preserve ageless wisdom regarding radical awakening and ancient wisdom on plane access, there are numerous other examples of cultures and communities that have helped to preserve teachings on shifts in identification, the cultivation of multidimensional relationship, and penetration into subtle planes (ancient wisdom).

One of the questions we must therefore ask ourselves, when coming into contact with evolutionary, ancient, and ageless wisdom on the nature of Reality and the kosmos, is what altitude of radical awakening, identification, developed relationship, plane access and structural intelligence was held by the people doing the 'knowing' and holding the 'wisdom'? The truths and deep wisdom of the kosmos are differentially disclosed at various altitudes of development in the two vectors. In other words, not only is reality enacted according to structural development but reality is also brought forth, at least to a certain degree, by development along the other lines of the evolutionary awakening vector, and by more and more humans radically awakening. As the mystics and spiritual practitioners of the ages have attested, on the planes of the soul (higher mental, buddhic, and lower atmic), for example, space and time function by fundamentally different rules. There, the radiant wisdom of Reality and the kosmos pervades the collective consciousness and energy-fields in a much more nakedly disclosed fashion. This is even more so on the monadic planes (higher atmic, monadic and logoic).

It is often true that the collective level of structural development along the intelligence sub-line of the evolutionary vector of those communities who cultivated extraordinary insights into ageless and ancient wisdom, was pre-rational and ethnocentric.[116] However, we would suggest that this does not necessarily invalidate the wisdom they held in relation to radical awakening and the other evolutionary sub-lines. What it *does* mean is that we must be careful in taking on their interpretations and contextual frameworks for that wisdom, both of which are functions of the structure of mental plane intelligence development that the people in question were operating through. Acknowledging these elements therefore allows us to make a clear distinction between what we understand as the difference between

---

116. And yet, it is important to remember that the level of structural development along this intelligence sub-line among the practitioners who actually had these reali- sations and brought through these teachings was often higher than pre-rational and ethnocentric.

'evolutionary', 'ancient', and 'ageless' wisdom, and it sets the stage for further dynamic synthesis as this book proceeds.

Earth is Eden

Chapter 15
Spiritual Exemplars
Jon and Dustin

Just as the various lineages and regions of the Earth have explored the territory of the two vectors with different areas of focus, so also have there been great spiritual teachers and masters who have been or who are powerful exemplars of a particular vector or evolutionary sub-line of spiritual development. In order to help clarify the two vectors and four evolutionary sub-lines introduced thus far, and their contribution to understanding the different paths of the One Way, we provide several examples of spiritual teachers and heroes who represent exemplars of each aspect of human potential.

Radical Awakening

In Chapter 3, we proposed that teachings on vantage point development (radical awakening) have been pioneered with the most granularity in the Eastern spiritual traditions. Accordingly, as our exemplar of radical awakening, we cite Gautama Buddha.

The historic Buddha is fantastic example of a great spiritual leader who taught his students how to deconstruct reality, vantage point to vantage point, to its naked Absolute Base. Repeatedly the Buddha taught his students to consider the nature of Reality, to see through their false assumptions about the apparent solidity of self and world, and to cut through dualistic perception so as to reveal the true nature of things. Awakening to the empty nature of everything within the manifest universe – all levels of relative vantage point, identification, relationship, plane access and intelligence – so as to reveal Ultimate

Reality as buddha nature, the Awakened Awareness that is emanating itself as all creation, stood as the foundation of his many teachings.[117]

A central foundation of the entirety of the Buddha's teaching is Emptiness (sunyata) Sometimes people have found the Buddha's teaching on emptiness somewhat nihilistic, negative, dry, and depressing even. For some, including such great saints as Sri Aurobindo, this teaching has felt a far cry away from the seemingly all-positive and ultimately embracing Vedantic notions on the unity of Atman and Brahman – that the deepest self of every being is always already one with the Absolute Self.

Understanding this requires us to tease apart the various different ways in which emptiness is actually understood in the Buddhist teachings. Two of the most important of these relate to it referring to either a) the impermanent, conditioned, and thus non-ultimate nature of the entirety of relative reality, or b) Ultimate Reality itself.

The use of the term to refer to the first of these is the sharp knife used in Buddhist teaching and practice to cut through our fixation on impermanent, conditioned, finite and conceptually designated objects in relative reality. This includes all perceptions, sensations, feelings, thoughts, personality, time, space and individual consciousness – all relative vantage points. As these are seen to be fleeting, ever-changing, finite and relative representations and constructions of reality rather than reality itself, our fixations upon them melt away. As the finest fixations at the subtlest levels of mind are seen through, the knot of self-contraction is undone, and our primordial condition is revealed as always already the unconditioned Awareness that is the Ground of Being.

The use of the term emptiness to refer to the second of these, Ultimate Reality itself, should not be seen to point to the absolute nature of things as some nihilistic nothingness. Rather, as the highest and most profound schools of Buddhism tell us – Dzogchen, Mahamudra, Great Madhyamaka and Zen – it points to the true nature of reality as an infinite, groundless ground of pure openness that is also unconditioned consciousness, unending light and dynamic energy emanating as all things indivisibly.

The Buddha taught that this pure openness is the true nature of every being. In Buddhism it is called buddha nature. And in a way that replicates the Vedantic teaching on Ultimate Reality as the Absolute Self, in some of his teachings, such as the Mahaparinirvana Sutra, the

---

117. The fact that Gautama Buddha demonstrated radical awakening from and without any attempt to transcend the personality altitude of identification is the reason why it is said in the Trans-Himalayan teachings that he embodied the perfected fulfilment of the 3rd aspect of personality/mind/form.

Buddha even goes so far as to describe buddha nature in terms of being the "true self" of every individual. This is despite the heavy focus in so much of the rest of his teaching on anatman, or the non-reality of any individual self.

This should not be seen to represent any conflict in the Buddha's teaching, but actually points to the clarification he brought to our ability to see, understand and experientially differentiate between the conditioned, finite, impermanent self-structure of the individual, and the radically unconditioned, infinite and unchanging Ultimate Self that is their true nature.

This clarification had particular significance for the time in which the Buddha was incarnated. As the recently translated book, *The Atman-Brahman in Ancient Buddhism*,[118] by the esteemed scholar, Kamaleswar Bhattacharya, shows, the way that the atman or true self of a person was understood during the time of the Buddha was much less in terms of a universal, all-pervasive impersonal Absolute and more in terms of a permanent personal atman, or self. The prominent notion at the time was that the fundamental self of every individual being was distinct; and permanent, indestructible, absolute and ultimate in its distinctness. This is a doctrine that can still be seen in Jain scriptures, for instance.

The Buddha did not deny the reality of a universal all-pervasive atman – a single divine essence that is one with all Reality for all beings; the one at the root of the many. Indeed when asked about it, it is recounted that he remained silent, unwilling to either affirm or deny. Rather, he denied the ultimate reality of a personal atman – a personal self, and thus showed the relative world and self, all the way through all vantage points, as impermanent, dependently arising, finite, conceptually designated constructions of experience, and thus non-ultimate; empty by any other name.

In much the same spirit as the "Neti Neti" approach employed by the Vedic Rishis of the Upanishads, the Buddha's teaching on emptiness illuminated the Absolute Reality that is left when all obstructions to its recognition have been cut through. Indeed, the three tiers of his teaching that went on to form the three turnings of the Wheel of Dharma – the truth of suffering and the emptiness of self (as focussed on most strongly in Theravada), the ideal of the bodhisattva and the emptiness of phenomena (as focussed on most strongly in Mahayana), and the true nature of all things as buddha nature (as focussed on most strongly in Vajrayana) – elucidated the nature of Ultimate Reality, and

---

118. Bhattacharya, 2015

the path of radical awakening in a way that was, and still remains, ground breaking in its granularity, nuance and finesse.[119]

All of this points to and just how profoundly all-embracing, clear and practical the Buddha's teaching was on the nature of Ultimate Reality and the path of radical awakening. Of course, it is important to note his teaching involves a path of opening to Non-dual Reality

---

119. Some scholars, including Ken Wilber, have questioned the degree to which the Buddha's realization penetrated all the way into Non-dual Reality, suggesting that the primary focus on recognising the empty, conditioned and finite nature of all relative reality that is the central focus of many presentations of the Buddha's teaching points more to a Causal level of vantage point opening. Causal realization opens the practitioner to the identification of their consciousness with the unmanifest formlessness out of which all manifest form at the Gross and Subtle states/phases of creation arises. Causal realization, when strong and abiding, is often characterised by such a deep absorption in this unmanifest formlessness that all relative form in the Subtle and Gross states falls away and is seen as illusory. Prior to the time of the Buddha such a realization was often equated with enlightenment, as can be seen in the Samkhya philosophy that formed the core of Patanjali's Yoga Sutras. The fundamental duality that this realization embodies though – unmanifest formlessness and illusory form – hinges on the last relative vantage point of individual consciousness not having been released so as to reveal the naked Awake Light that is the root and essence of all states/phases of Reality – Causal, Subtle and Gross. As long as this is the case, not only is the last veil obscuring Awakened Awareness not seen through, but having seen the entirety of relative reality as it is dualistically enacted through all lower vantage points as ultimately illusory, it is not then reintegrated from radical awakening as the radiant expression of Awakened Awareness.

That the Buddha's realization and teaching did take Awakened Awareness as its Primordial Ground can be demonstrated by his explicit teaching about it, as can be found in such Tathagatagarba Sutras as the Mahaparanirvana Sutra and the Infinite Life Sutra. Of course there are those who might ask the question of whether the teachings embodied in these sutras really were the teachings of Gautama Buddha or whether they were developed later in the evolution of the Buddhist tradition. Such a position would see the only reliable sources of what the Buddha truly taught as found in the Theravada School, which was the earliest formulation of the Buddha's teaching and that came before the Mahayana, which developed as a distinct tradition around 2000 years ago, and the Vajrayana, which developed around 1000 AD.

As is often missed in the conventional taxonomy of the Buddha's teaching however, there is evidence that the lack of focus on those points of teaching that became central for the Mahayana and Vajrayana – the emptiness of phenomena, the ideal of the bodhisattva, and buddha nature (Awakened Awareness) – in the First Turning of the Wheel of Dharma teachings (Theravada) is not a true representation of what the Buddha taught. Rather, it is an expression of which aspects of his teaching became most emphasized by the dominant sects in the earliest genesis of the Buddhist tradition. For more in this, see the work of Dr. Tony Page on the Mahaparinirvana Sutra (http://www.nirvanasutra.net/), and John Reyonld's book, The Golden Letters, pages 283-286.

according to a 1st person perspective. That is, it entails engaging the path of radical awakening through an exploration and revelation of the true nature of one's own 1st person consciousness. As we have said previously though, radical awakening can also be opened up according to 2nd person and 3rd perspectives. Exemplars of radical awakening opened up according to a 2nd person perspective would be the Sufi mystics, Ibn Arabi or Jalāl ad-Dīn Rumi, for instance. An exemplar of radical awakening opened up according to a 3rd person perspective is the Taoist sage, Lao Tze.

Exemplars of Evolutionary Awakening: Identification

In the Trans-Himalayan teachings, it is understood that one of the most profound exemplars of the shift of identification from personality to soul to monadic altitudes is Christ.[120] Each of these transformational shifts can be seen as related to specific events in his life.[121] According to our current consideration, these began with his baptism by John the Baptist, where he was subtly initiated into realization of his spiritual work and after which he began teaching. In the transfiguration experience on the mountain top, Christ's identification fully shifted from the personality into the soul and was expressed in his realized trans-incarnational lineage connection with the Masters, Moses and Elijah. The next shift on his path from soul into monadic identification can be seen in his experience leading up toward the crucifixion. This began in the Garden of Gethsemane, where the Trans-Himalayan teachings understand that Christ's surrender into the divine Will and envisioned Purpose of the planetary Logos of the Earth was opened up with the mantram "Father, not my Will, but thine be done". Later, his monadic identification with and as the Life of that Being was stabilized with the affirmation: "I and the Father are one".

Indeed, in the Trans-Himalayan teachings, it is understood that Christ's use of the term "Father" referred specifically to his unfolding relationship with this vastly-evolved being that stands as the planetary Logos of the Earth. The opening up of such a relationship, whilst retaining identification with the entirety of both humanity *and* Hierarchy, is understood as unprecedented in our planetary history

---

120. This is an interesting point of departure from conventional Integral Theory, where the narrative of Christ's path has been primarily considered from a radical awakening perspective rather than in terms of his evolutionary development.

121. And these can be seen to relate to the initiations we discuss in the following chapter.

according to the Trans-Himalayan teachings. In this connection, Christ can be seen as one of the foremost exemplars of the relationship line of the evolutionary vector also.

Christ additionally serves as a clear example of the unfoldment of lines of intelligence extended right through the personality and into the soul altitude of identification. His lines of intelligence particular to the personality altitude (moral, emotional, cognitive, interpersonal) were all clearly operating at post-conventional levels around the time of the transfiguration. Subsequent to the transfiguration, we see a profound employment and use of the intelligences specific to the soul planes— intuitive mind, love, telepathy, and innate healing capacity.

In the last three years of his life, Christ remained an incredible example of a being engaging this shift from soul into monadic Life—a radically awake, living embodiment of pure love and surrender into the spontaneously moving Will of God. It is this divine Will that is shown in the Gospel as pouring through his heart more and more fully from his cultivated identification and relationship with Shamballa and the planetary Logos of the Earth, and to which his life was committed regardless of consequence.

With the shift from soul into monadic altitudes, he pierced into the sphere of planetary Will, Purpose and Life that characterizes Shamballa. As we have described before, in the monadic altitudes the Nondual and Causal states of Absolute Reality naturally shine forth, and so as this shift occurs, the sword of radical awakening cleaves through every final remaining trace of self-contraction and misidentification with a relative self ("Before Abraham was, I AM"). All that is left is Absolute Reality itself expressing effortlessly in the apparent form of a being identified as the incarnated Life, Love and Light of the planetary Logos within whose body they find their place. Again, we see this in the life of Christ in his ever-deepening identification with the Will, Purpose and Life of the Father, or planetary Logos of the Earth, leading up to the crucifixion. Indeed, Djwhal Khul, in his work with Alice Bailey, suggested that what is perhaps the clearest exposition of this process of monadic identification and relationship with the planetary Logos, in all the world's scriptures, can be found in John 17: 20-26. In the words of Djwhal Khul:

> There is no other passage in the literature of the world which has exactly the same quality. Oneness, unity, synthesis and identification exist today as words related to consciousness and as expressing what is at present unattainable to the mass of men. This manifesto or declaration of the Christ constitutes the first attempt to convey reaction to contact with Shamballa, and can be correctly interpreted only by initiates of some standing and

experience. A concept of unity, leading to cooperation, to impersonality, to group work and to realization, plus a growing absorption in the Plan are some of the terms which can be used to express soul awareness in relation to the Hierarchy. These reactions to the united Ashrams which constitute the Hierarchy are steadily increasing and are beneficently conditioning the consciousness of the leading members of the forefront of the human wave at present in process of evolution.

Beyond this state of awareness there lies a state of being which is as far removed from the consciousness of Members of the Hierarchy as that is, in its turn, removed from the consciousness of the mass of men. Endeavour to grasp this, even if your brain and your power to formulate thought rejects the possibility of this exalted livingness. Be not discouraged at this inability to understand; remember that this state of being embraces the goal towards which the Masters strive, and which the Christ Himself is only now attaining.[122]

Here, in John Chapter 17, as we understand it, Christ prays in heart-communion and attuned identification with the Father, or planetary Logos of the Earth, and with reference to his disciples:

I do not pray for these alone, but also for those who shall believe in me through their word,

That they may be one, as You, Father, are in me, and I in You; that they may also be one in us, that the world may believe that you sent me,

And the glory which You gave me I have given them, that they may be one just as we are one:

I in them, and You in me, that they may be made perfect in one, and that the world may know that You have sent me, and have loved them as You have loved me.

Father, I desire that they also whom You gave me may be with me where I am, that they may behold my glory which You have given me; for You loved me before the foundation of the world.

Oh righteous Father! The world has not known You, but I have known You: and these have known You that you sent me.

---

122. Bailey, 1960, p. 173.

And I have declared to them Your name, and will declare it, that the love with which You loved me may be in them, and I in them.[123]

Relationship

Bruce Lyon is the founder of Shamballa School, and it is through his consciousness-based relationship with the Master Djwhal Khul—who is the same soul plane Master that both Helena Blavatsky and Alice Bailey worked with in different ways—that a new series of teachings within the Trans-Himalayan movement have begun to emerge. In the quotation below, Bruce explores the processes of contact with such a being, and his own first-person experience of it:

> In the year 2000 I began a remarkable relationship in consciousness with a Tibetan spiritual teacher named Djwhal Khul, or DK. I was familiar with his work in the Trans-Himalayan tradition and was a student of his writings via Alice Bailey.
>
> Like all relationships there was a 'dating period' where we got acquainted with each other, or rather reacquainted, for it appeared we had worked together before and had a pre-incarnational contract. The early stages of our relationship were very difficult for me as I struggled with my core self-worth wound in the presence of so powerful a current of love-wisdom. It was also challenging to validate the relationship as it was occurring wholly on soul and monadic levels. On the other hand I did not have to deal with the jarring dissonance that often occurs when relating to the personal life of an embodied teacher.
>
> I approached the connection like a scientist—making a regular time and place for our meetings, recording any impressions together with my reactions. I also noted the effect of the energetic impression on my own system and on those around me. The relationship was not so much a meeting of minds but a soul communion—it was as if our subtle selves came together in what Rumi would call a 'sohbet' and my awareness was stimulated, opened and flooded with insights as a result. After a period of

---

123. John, 17: 20-26

testing I agreed to collaborate in the writing of a series of books. From the perspective of DK, it allowed him to begin to release the next phase of his teaching work, and my journey of awakening was greatly accelerated as energy and information of high vibration moved through my system.

The essence of the teaching, and the effect of the relationship, was to facilitate the opening of consciousness into its core and reveal the Life or Presence of Divinity that resides there as the nondual root of awareness and form alike. Earlier teachings had been about expanding consciousness and developing the capacity to penetrate into (and be penetrated by) more and more subtle realms of experience. This teaching directed the awareness back upon itself, opening the soul to a direct experience of the 'living word'.

Essentially it was a transmission of freedom from a being who was already living in that experience and it was my work to clothe this transmission in words and concepts that would both veil and reveal it to others. Like all good teachers this Master eventually directed my attention away from my relationship with him and focused it on the font welling up in the core of my own being.

I am incredibly grateful for the experience—not only for the grounding in Being that has resulted, but also for the awakening in my heart and mind of the direct experience that humanity is not alone. Great souls and whole kingdoms of presence, power, love and intelligence surround us and interpenetrate our realities, waiting for our maturity and cooperation in a vast cosmic drama. Life on this glorious Earth, with all its diversity, beauty, suffering and joy, is a great mystery that is always revealing more of its hidden depths to those who inquire the way. It is my great pleasure to be part of that inquiry.[124]

Plane Access

Vision into physical, etheric, and astral planes is a signature capacity developed in many indigenous and shamanistic traditions. In fact, this

---

124. Written for this book.

may be the most ancient form of spirituality on our planet. We honor the incredible capacity for vision and penetration that some of these individuals truly have. Below, we offer three exemplars of this vector: two indigenous shamans and one Westerner.

First, we look to Don Jose Campos, a Peruvian shaman. Don Jose writes of what is like to begin opening vision into the etheric and astral planes. According to his own path, vision began to open as a result of working with plant medicines, including Ayahausca. He writes:

> I will begin to explain by telling you about a vision I had with one of my teachers. Here in the jungle, the *curanderos* or *vegetalistas* who drink the plants do so partly to acquire their power. This power or force is called *mariri*. It means the power of the plants you have taken. In an Ayahuasca ceremony with one of my teachers, I could see the power of the plants inside his body. So with his song, his *icaro*, he began to call forth the power, not only of the plants, but of animals as well. He invoked the power of the *otorongo*, the jaguar. In the vision I was having of him, I could see him place the spirit of the jaguar in the person he was healing, to protect him. This spirit or energy was transmitted through his *icaro*. It is here that we step into the world of shamanic forces because the shaman has the power to invoke these energies, these spirits.[125]

He goes on to affirm again the reality of the capacity to see with subtle vision:

> One can clearly see the power of protection. It was not just me. The person involved also saw the protection. He came out of the ceremony healed and stronger and with a lot more energy. To me, that is shamanic. It is through drinking the plants that one can obtain this force and the transfer of this force is shamanic. Of course, I have the vision because I drink the plants.

An analogous capacity for subtle vision is cultivated in other indigenous cultures as well. Malidoma Patrice Somé, an initiated elder of the Dagara people of West Africa, offers us a perfect example of someone representing a culture in which access to inner planes (physical, etheric, astral) is fully integrated into social structure of society. Malidoma explains that within his culture there is "no distinction between the natural and supernatural world." The world of the supernatural—or what we would call the etheric and astral

---

125. Campos, Grob, & Roman, 2011, p. 3-4.

planes—is a realm of various entities, ancestors and a whole world of other energies and creatures that one can learn to interact with. Malidoma writes of how vision into the astral plane is cultivated and contextualized among the Dagara people. He explains:

> The Dagara believe that contact with the otherworld is always deeply transformational. To successfully deal with it, one should be fully mature. Unfortunately, the otherworld does not discriminate between children and adults, seeing us all as fully-grown souls. Mothers fear children opening up to the otherworld too soon, because when this happens, they lose them. A child who is continually exposed to the otherworld will begin to remember his or her mission in life too early. In such cases, a child must be initiated pre-maturely.[126]

Because indigenous cultures often tend not to record stories and histories in written form in the same way that Western cultures do, it serves us to bring in one Western exemplar of plane access where our descriptions can be even more precise. In this case we look at the exemplar Edgar Cayce. Cayce provides a fascinating case of an individual who gained a substantial degree of vision into multiple planes of reality; because thousands of his cases and readings are recorded, there is a vast amount of data to support his capacities that help to fill out this vector.

Cayce repeatedly demonstrated a capacity for remote viewing medical diagnosis of physical ailments. We use this example to point out a particular capacity for access to planes (in this case physical-etheric). To give a sense for his accuracy of vision, we note a survey that was conducted by a journalist five years after his death. The survey examined the reports of eleven doctors, all of who worked closely with Cayce's capacity for vision, and the results of which survey directly support Cayce's capacity for subtle sight into the physical body.

> "A doctor in Bronxville, New York, evaluated Cayce's diagnoses as 100 percent correct in the twelve cases he had treated. After treating twenty persons who had received readings, a Detroit physician estimated the accuracy of diagnosis at eighty to ninety percent... In Albany, New York, a cooperating physician stated that all five patients he had seen had received correct diagnosis." [127]

---

126. Somé, 1995, p. 19.

127. Johnson, 1998, p. 17

Clearly, Cayce had a well-developed capacity for vision into various aspects of the physical and etheric plane.[128]

Lines of Intelligence (The Cognitive Line of Development)

Some of the hallmarks of a high level of structural development are the capacity for complex reasoning, perspective taking, and various expressions of knowledge synthesis through sophisticated forms of pattern recognition. At its highest levels, pattern recognition gives way to direct perception of "thinking, seeing, feeling, witnessing and being" complex systemic wholes.[129] At these most advanced levels it is as if information begins to be received in chunked wholes, already fully-formed and complete, only later to be unpacked.

As we examine the small sample of highly developed human beings in this dimension of structural development, one particular individual stands out as exemplary. Integral philosopher Ken Wilber, the grand synthesizing theorist of our time, provides a fitting case study to demonstrate how the highest levels of structural development look and feel from the inside. Wilber's post-metaphysical orientations and the system of analysis known as Integral Methodological Pluralism (IMP) represent extraordinary expressions of his particular level of structural development. As a paramount aspect of his methodologies and theory, these highest levels of structural development use an unprecedented capacity for perspective taking to honor successfully the truths of a multitude of alternative methodologies already in existence while simultaneously integrating them into a larger, fully coordinated, holistic matrix.

An integral view, according to Wilber, is recognizable first and foremost by its degree of *comprehensiveness, inclusiveness, non-marginalization, and capacity for embrace*.[130] Wilber explains it in this way:

> Integral approaches to any field attempt to be exactly that: to include as many perspectives, styles, and

---

128. Interestingly, Cayce might also be an example of access to the Causel state. A Casual level of realization is beyond limitation of both time and space. It is reported that Cayce was able to access information about the distant past that was then later verified by scientific study.

129. See Wilber's Audio Interview on "Patterns of Wholeness: A Firsthand Account of the Highest Structures of Consciousness" from Core Integral Loft Series.

130. Visser, 2003, p. xii-xiii.

methodologies as possible within a coherent view of the topic. In a certain sense, integral approaches are "meta-paradigms," or ways to draw together an already existing number of separate paradigms into an interrelated network of approaches that are mutually enriching.[131]

Wilber himself has moved beyond the beginning stages of an integral orientation (the turquoise structure previously mentioned) and finds himself at its more advanced iterations. At these higher stages, a certain degree of vantage point development and deeper level of identification becomes a requirement for higher structural growth to stabilize. Wilber calls these higher integral stages Vision Logic, Illumined Mind, Intuitive Mind, Overmind, and Supermind. His descriptions follow the original names given to these higher stages by the Indian sage Aurobindo, but he adds to them quite extensively given the latest research within developmental studies.

Explaining his own experience, Wilber articulates what it is like to process information at these higher levels of structural development: "Vision logic is thinking wholes, Illumined Mind is seeing wholes, Intuitive Mind is intuiting [feeling] wholes, Overmind is witnessing wholes, and Supermind is being whole".

According to Wilber's own descriptions, he tends to spend most of his time in these highest two integral stages. He explains:

> Overmind and Supermind is in a sense where I spiritually hangout. They are also the two modes that I use to check AQAL Theory [Integral Theory]. I use these to check AQAL theory against everything that's arising. I have some sense that AQAL theory is adequate to everything that is arising. When I write about AQAL theory I translate it into vision logic and most of the writing I do is from the level of vision logic. And that's simply because it's the highest level I can count on most individuals being able to follow without being in over their heads. Then, Illumined Mind and Intuitive Mind are the two modes that are most often used for Integral theoretical work. When I'm looking at a problem [or] when I'm trying to deal with a particular issue; when I'm trying to figure it out, then I'll use one of those two modes of knowing. As I said, I generally check it with Overmind and Supermind, but actually do the work with Illumined Mind and Intuitive Mind. The reason is that between the two of them they cover gross, subtle, and causal [events/objects], and so I can rest assured that

---

131. Visser, 2003, p. xii-xiii.

I'm covering most of the important points that need to be covered. And then of course I'll translate that into Vision-Logic and then periodically check it against Overmind and Supermind. This checking is something that I can't do, for example with Big Mind [Awakened Awareness, alone]. Big Mind [Awakened Awareness] is just the state of Empty Awareness and it can occur and be experienced at virtually any level. So it doesn't really have any content. It doesn't really have anything to tell me. And that's one of the big differences between Big Mind and Supermind. Supermind is Big Mind *plus* all of the structures that have emerged at first, second, and third tier. It's an important difference. It's why both Supermind and Big Mind are important for realization, for genuinely waking up.[132]

Wilber further considers the higher reaches of structural development:

"Is there something higher than Supermind? Probably. When I operate in Supermind… it doesn't seem like you can't go any higher. It seems like you can. So what's higher than that? I have absolutely no idea. We'll just have to wait and see."

Integral Theory suggests that as higher structure-stages of evolution unfold and stabilize, these views will become commonplace; it is inspiring to imagine what that might look like and the ways in which we human beings will then relate to our understanding of the world, and each other.

---

132. See Audio Interview by Wilber "Patterns of Wholeness: A First hand Account of the Highest Structures of Consciousness"

Chapter 16
Relationships Between the Vectors: From Anatomy to Physiology
Jon

Up until this point, the flavor of our vector exploration may have felt a little fragmented, as if we are considering each vector and their associated lines as isolated aspects of unfoldment. This is not so. Although the distinctions between the vectors and lines are important to help us a) know where we are in time and space with our own developmental process, and b) to understand more deeply the various gifts that are brought to the world table by the different spiritual traditions of the Earth, it also seems true that at a certain point in spiritual advancement, the distinctions become less important as the streams pour into one river. At that point we suspect that the rune symbols of kosmic truth and process are so clearly perceived in every particle of the manifest universe, from the bark of trees and the movements of water to the folds and ripples of subtle energy-matter on the inner planes, that journeying on the Path becomes self-evident.

In this chapter, we thus turn our attention to the much more fluidic and interpenetrative relationships between the two vectors that characterize the reality of the Path. We will not explore every combination of relationship between the vectors and various-lines in each, but will instead only explore those relationships that seem to have greatest relevance to each other, and to an individual's growth. May these distinctions and the relationships that we describe between the vectors be helpful for those looking for articulations of the further reaches of human possibility.

Radical Awakening and the Evolutionary Line of Identification

Ultimately, for the purposes of clear differentiation in our work, the radical awakening vector refers to the *deconstruction* of the entire kosmic spectrum of relative reality to its naked Absolute Essence. The identification line of the evolutionary awakening vector refers to the infinite series of transfigurations of identity *within* the spectrum of the manifest kosmos. The sense of identification is always relative though, no matter to what scales of time and space in the kosmos it expands in its ever-widening identification. Ultimately there is only ever the Open Sky of Absolute Reality arising as patterns of energy (the bodies of all levels), rays of light (consciousness on all levels) and bolts of lightning (emanations of monadic Life on all levels). Radical awakening continually reveals this Reality in its unobstructed nakedness as empty, infinite Primordial Being.

This relates to the horizontal and vertical definitions of spirit and matter given by Djwhal Khul and Bruce Lyon in their book, *Occult Cosmology*. There they make the point that spirit and matter, when viewed dualistically, can be understood in both a horizontal (radical) and vertical (evolutionary) sense. This can be seen in the figure below. Horizontally, spirit is the unborn and ever-present root of all subjectivity—the Boundless Immutable Principle of Awakened Awareness that transcends, includes and emanates as the entire spectrum of kosmic planes. Vertically, within the relative universe, spirit is the Fire of Universal Life—the One Universal Monad that we are in the deepest kosmic altitudes of our identification line, and that is revealed in ever wider and more inclusive forms as identification is shifted into the monadic bodymind and beyond. In the words of Djwhal Khul and Bruce Lyon, "Spirit is Matter at its highest vibration while Matter is Spirit at its lowest vibration AND Spirit is Matter at its most subjective while Matter is Spirit at its most objective."[133]

---

133. Lyon, 2010, p. 227.

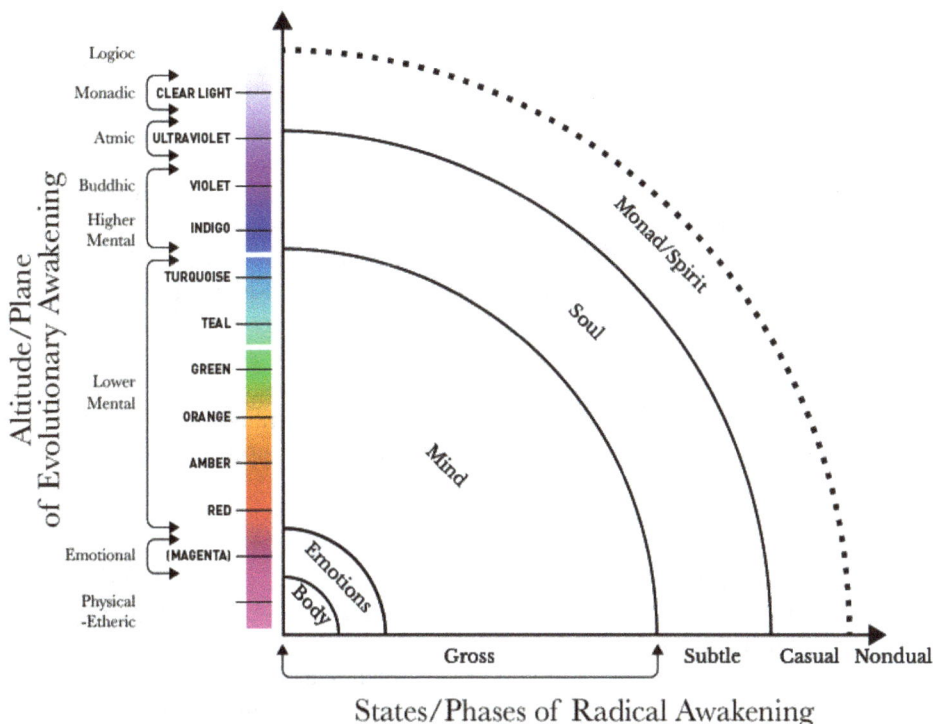

Figure 15. Chart showing the interpenetration of radical and evolutionary vectors resulting in the arising of progressive levels of identification.

As we pointed out in Chapter 9, both the most recent iterations of Integral Theory and the Trans-Himalayan teachings posit that each of the evolutionary structures partakes in a particular state or vantage point of Awakened Awareness's involution into Form. Specifically, we suggested that on the monadic planes the Nondual and Causal states of the Absolute shine forth with greatest clarity; on the soul planes the Subtle state of the Absolute is most prevalent; and on the personality planes the Gross state of the Absolute is most obvious. As we discussed there, this has deep implications for the way we understand the relationship between radical awakening and the various depths of evolutionary identification on those particular planes. Specifically, with this perspective in place we can see that the monadic depth of identification remains ever awake to and as Absolute Nondual Reality and the Infinite Abyss of Causal formlessness. We can see that the soul depth of identification, owing to the perpetual shining of the Subtle state of the Absolute on the soul planes, need only glance into the nature of its own awareness for full radical awakening to unfold. We can see that though radical awakening remains available to all selves operating on the personality planes, the prevalence in vision of the Gross state of the Absolute tends to obscure the ever-present Truth of Ultimate Reality.

These points have some compelling implications for how we understand the interactions between radical and evolutionary awakening processes. Specifically, we can see that owing to the Nondual and Causal states of the Absolute being most nakedly pervasive on the highest three planes, an evolutionary shift of one's identification into the monadic bodymind (clear light altitude) *necessarily entails* stabilized Causal into Nondual radical awakening. Similarly, since the Subtle state of the Absolute shines so pervasively on the soul planes (indigo-ultraviolet altitude), we can see that a shift of identification into soul *necessarily entails* radical awakening at least to the Subtle state of the Absolute—an infinite field of light, sound and energy. Owing to the fact that the Gross state of the Absolute is most prevalent to our vision on the personality planes (infrared-turquoise altitude), those whose identification is rooted here—which encompasses the majority of humanity—will have awareness of the Gross state of the Absolute, largely in its physical plane expression—the material universe. Since its Subtle, Causal and Nondual levels are not as obvious to relative vision on the personality planes as they are on soul and monadic planes however, the majority of human beings will not experience the continuous recognition of the Absolute Reality of the Great Perfection through all phases—Gross, Subtle, Causal and Nondual.

As the radical awakening traditions, such as Dzogchen, Mahamudra, Zen, Vedanta, Kashmir Shaivism, Sufism, and Bon teach, however, radical awakening to the One Nondual Reality is ever available on the personality planes (specifically the mental plane). Radical awakening requires no further evolutionary awakening than a stable sense of self, as said earlier by Dustin. This has allowed the radical awakening traditions to have successfully taught this path without any attempt to shift people's center of gravity in the self any higher in altitude than the levels of personality, for millennia.

Just as evolutionary awakening into deeper and deeper altitudes of identification (from personality to soul to monad) necessarily evokes radical awakening to the Nondual Absolute (via Gross to Subtle to Causal states), radical awakening, as it is stabilized, opens the door and lubricates the path of evolutionary awakening. This is naturally the case owing to radical awakening entailing the falling away of resistance to Ultimate Reality's own display to itself. As this unfolds, the Clear Light of Source becomes increasingly able to express through each level of relative identification on the kosmic physical plane simultaneously.

That is, Awakened Awareness then expresses freely through monad, soul and personality all at once. Total continuity of Reality realization is then established as the basis of all three levels of identification, yet with Reality operating seemingly independently through each.

In this way, radical awakening opens the path of evolutionary awakening for shifts of identification into deeper and deeper levels of spontaneously arising creativity, love, and power in the unfoldment of kosmic Purpose. As this unfolds, and Awakened Awareness recognizes itself through a human being, their relative sense of self dissolves into Clear Light—an empty sky through which patterns of energy, rays of light and bolts of lightning naturally arise through the apparent form of 'a master'. In the words of Djwhal Khul when speaking to some of his students:

> As your Master I must tell you that I do not exist. I am a coating, a covering on that One Life that you can recognize. The difference between us is that I know that I do not exist in any real way that is separate from the One Life itself and you do not yet. I am only your Master to the exact degree you are my servant. When you surrender to My will you are merely surrendering illusion to a wiser and more lively illusion that is present within your Self. In that surrender you release your identification with that small and temporary home (now a prison to you) called the human soul and take possession of a roomier mansion. The difference is now that you know that mansion is part of the 'father's house' and you as the monad are not separate from me or Him or the One Life itself.[134]

In fact, we would go so far as to say that radical awakening has an electrifying effect on *all the lines of the evolutionary awakening vector*. This means that if one stabilizes realization in the Nondual Absolute as the ever-present Base of one's awareness, heart and body, this naturally leads to the unfolding of deeper altitudes of identification. In this process the egg-shell of relative identity continues to crack open to reveal the path of monadic emanation along which we *came in*. This evokes the opening up of relationship with other Self-reflections of the Absolute on progressively celestial levels. It calls forth our opening up of access to progressively subtler realms of experience. And it summons the rippling forth of ever-deeper developmental expressions of kosmic

---

134. Three Jewels Transmissions, 2006.

intelligence and creativity. In the words of Bruce Lyon, "Radical awakening is like pouring gasoline on all the other fires".[135]

As we have hinted already and will go to explore more fully in the next section, the shifts along the evolutionary awakening line of identification do not end with the monadic bodymind. Rather, from here, and in post-mastery stages of maturation, the monadic essence is able to be abstracted onto higher kosmic planes, such as the kosmic astral, kosmic mental, kosmic buddhic, etc. There, the mastery it has uncovered and cultivated during its journey may be employed to incarnate through planets, solar systems, and eventually such massive bodies as galaxies and universes.

In this respect, though the kosmic evolutionary path for all beings is one, the Absolute Reality of Awakened Awareness that is realized in radical awakening expresses *uniquely* through every monadic essence. It is thus the destiny of every being to give a perfectly unique expression to their awakening through all altitudes, even as their path stretches out into the kosmos—one so precious that it has never been expressed before and never will be again. As identification shifts deeper and higher, and continuity in consciousness is established between the various bodyminds and altitudes of identification, the power of love, inclusivity, wisdom, insight, and kosmic intelligence through which that unique expression of the Absolute is enacted, simply catch fire.

In relation to these advanced stages of unfoldment, we suggest that the combination of radical awakening and shifts of identification into soul and monadic altitudes underlies certain of the miraculous phenomena described in the world's esoteric lineages and scriptures. One example is the attainment of the Rainbow Body or Great Transference as discussed by such Dzogchen Masters as Garab Dorje, Manjushrimitra and Padmasambhava, and the body of light/ resurrection of Christ. Additionally, we would suggest that the power of transmission capacity among spiritual teachers and masters is a function of the extent of parallel awakening along these two lines.

Interaction and Influences between Lines of Evolutionary Awakening: Identification and Relationship

Shifts in identification and relationship go hand in hand with each other, and this is closely tied in with plane access too. Specifically, one

---

135. Personal communication.

route by which shifts in identification can occur (from personality to soul to monadic bodymind), first as temporary peak experiences and then increasingly stabilized, is through the energetic empowerment transmitted by beings on higher planes with whom we have opened up relationship. This is a foundational point for understanding the path of initiation, which we will go onto describe later. Briefly put, one perspective on initiation is that it involves beings on subtler, higher planes giving a hand up, as it were, to those on denser, lower planes. This allows the purpose being transmitted by those on higher planes to be served and expressed more effectively by those on lower planes. It is this transmission of purpose that empowers transitions in identification for those whose altitude resides on lower planes, as their very identity, consciousness and bodies are charged with the frequency of higher altitudes. As the path unfolds, this eventually allows entry into whole new spheres of relationship with all kingdoms on all planes. This is why it is understood in the Trans-Himalayan teachings that one mode of deepening identification progressively into the monadic bodymind is through contact with the master who stands at the core of the ray ashram corresponding to the ray of one's monad (1st ray ashram for a 1st ray monad, for instance).

In the context of multidimensional planetary service, this is not a one-way deal, however, with only the 'climbing self' who is given the energetic 'hand up' benefiting. Rather, it is through beings', such as masters', transmission to those on denser planes that the former are able to incarnate through and/or manifest their projects of planetary service, as was pointed to in Chapter 12.

These points have a number of implications. First, we can see that one can open up relationship with beings on planes on which we have not yet accessed our own altitudes of identification (opening up contact with beings working on the higher atmic plane of planetary will transmission, for instance, without having consciously contacted this altitude of our own monadic identification). This also can work in the reverse; that is, it is possible to deepen identification progressively with and into a particular bodymind (the soul bodymind, for instance) without having opened up contact with beings residing already on those levels. The one caveat that should be added here is that by doing so, one necessarily *opens the door* to such contact.

Second, we can begin to understand further the two sides to contact with beings abiding on higher planes—the masculine pole of *penetration*, and the feminine pole of *fertile receptivity*. The first is the movement that pierces into the energetic field of, for instance, a master within Hierarchy. The second is being available as the grounding point through which a master's particular transmission can find root in manifestation. When speaking about this process in greater, more kosmic-scale contexts, we can speak of groups, spheres of evolutionary life such as humanity, Hierarchy, Shamballa, or the single unified consciousness of planets as a whole, penetrating into kosmic planes. There, they are able to contact and open to being the fertile form through which kosmic Logoi, avatars, and the vastly awakened beings incarnating through planets, solar systems, stars, and galaxies can express on the denser levels of the kosmos. This process will necessarily unfold according to a 3rd person to 2nd person to 1st person sequence, which allows the contacted being to move from an 'it to a 'you', to the expressing 'I'. As a result of this process, the abiding altitude of identification of the entity who is incarnated through is naturally deepened, and there is a continuously unfolding wisdom that emerges on the nature of the multidimensional ecosystem of this planet and the kosmos.[136]

Another point to explore in this context is the relationships that begin to open up for a monadic essence once they begin to liberate their identification from the sheaths through which they have been manifesting on the kosmic physical plane (the personality, soul, and monadic bodyminds). Specifically, as an individual deepens their radical realization that their True Nature is not limited to the sheaths through which they have been expressing or the entire kosmos, they also come to recognize that those sheaths are not simply 3rd person configurations of energy-matter but are actually each animated by their own 1st person conscious Life. This relates to the Trans-Himalayan teaching that each of our sheaths is literally an embodied deva that is incarnating through the elementals that compose the sheath, and evolving through its increasing responsiveness to being directed by consciousness.

Breaking that down, it is understood that the monadic body, the soul body, the mental body, emotional body, etheric body, and physical body are all *living devas*, though of differing degrees of development. Specifically, those who are the central lives incarnating through the mental, emotional, etheric, and especially the physical bodies are understood to be devas that are pre self-conscious, whereas those composing the soul and monadic bodies are super-conscious. As such, they are also monadic essences—the dynamic energy of Awakened Awareness, yet in the case of the sheaths of the personality bodymind, they are Awakened Awareness operating through sheaths that are not sufficiently evolved to have the capacity for self-reflexivity, and thus to allow these devas to be able to radically awaken by *Self-recognizing* as Awakened Awareness.

---

136. Trans-Himalayan students will recognise that this is why it is understood that the Master Jesus was able to take the 4th initiation as a result of his surrender to being incarnated through by the Christ.

Human beings who *do* have this capacity and who are moving into the advanced stages of radical awakening therefore hold a profound, planetary scale responsibility as the potential liberators of all the devic beings composing their embodiment. That responsibility is to surrender more and more deeply into Nondual realization so that it utterly saturates all form, including the lives that are their physical, etheric, emotional and mental bodies, allowing each and all of these to participate fully in radical awakening and dissolve in total Self-Realization. Again, this process naturally unfolds in a 3rd to 2nd to 1st person sequence, with that which is initially deemed simply 3rd person objective form (such as the physical body) recognized as a 2nd person sentient life with whom relationship is possible, and then self-liberated as the One through the realizer's radical awakening to the 1st person Awakened Awareness that is arising as the kosmos.

In terms of those super-conscious devas who reside as our soul bodies, these devas are understood to be beings passed through the human stage in long, long ages past, whose natural altitude of abiding now is the kosmic astral plane. From the incredible light and love that they have now moved into, however, these beings chose to sacrifice their place in the heights to descend onto the subtlest levels of the mental plane where they reside as the conscious life of the soul lotus-body described earlier in Chapter 10. On these levels, they wrap their bodies of purest, stainless light around each monadic essence so as to provide a mirror of subtle form through which each might first come to know itself in self-consciousness, and then radically awaken to its True Nature.

This they do from our very birth as souls right through to liberation at the 4th initiation. At this stage of the path, the electric energy of monadic identification and the stabilized radical awakening it brings forth gain such fiery livingness that the illusion of our identification with the soul is energetically burned through and the experience of monadic Life, beyond self-reflective consciousness, is entered. At this time, the devic being of the soul body is released back to the kosmic point from which it originally came forth—the Heart of the Sun.

Identification and Plane Access

In a similar fashion to the identification and relationship lines just explored, these two lines, identification and plane access, are as closely linked as well. Once again we speak of the two sides of this process being a masculine penetration and a feminine receptivity. It is therefore possible for individuals and groups to penetrate into subtler planes than they yet have a corresponding altitude of identification on, so as to work with their redemptive, powerful, and loving energies and gain access to information contained in the fabric of their energetic fields. It is also possible for such individuals or groups to receive those energies and act as vehicles for their manifestation.

One planetary scale example of this is the role the Trans-Himalayan teachings understand to be humanity's destiny to play: as a planetary shaman, which was discussed in Chapter 12. This involves humanity both penetrating into the transformative and transfiguring energies and information held in the evolutionary life-spheres of Hierarchy and Shamballa, and receiving them in a manner that allows their revelatory channeling into the planetary geosphere, biosphere, and noosphere, and into the core of the Earth.

Another point to make when considering the relationship between identification and plane access is that shifts in the former and shifts in the latter are definitely linked, but not necessarily hooked. Specifically, there is a 'link' between shifts in the altitude of identification (personality, soul, monad) and penetration onto the corresponding planes for that altitude of the identification (see Table 1). What this means is that shifts in identification to a particular altitude, let's say into the soul (though it could be any other depth also), opens the door to full penetration and access to explore and journey on the soul planes, *but does not take you through it*. When one does 'walk' through it, there is then the possibility of opening up access to the shared energies and relationships of that particular plane, the information that is encoded into its living fabric and the opportunity to learn to magically direct and manipulate its energy-matter.

The reverse is also true. Increasing access to the energies of particular planes can definitely open the door to a progressive shift into a deeper altitude of identification, but it will not necessarily take you through it. For example, the capacity to invoke and contact the energy of the monadic plane from the personality or soul bodymind will, owing to the fact that those energies are so powerful and soaked through with the Reality of Nonduality, increasingly open up the capacity for identification with the monadic bodymind, as well as radical awakening.

Identification and the Lines of Intelligence

Most research conducted on self-development is enacted from an altitude of identification abiding in the personality (infrared-turquoise on the color scale). At this level, identity is often entirely entangled in the mind, and as such most research into the evolution of identification at the personality level does not differentiate it from growth through structures of intelligence.

Harvard Professor Robert Kegan's book, *Evolving Self,* is a perfect example of this type of conflation. In Kegan's work, the self's identification moves through various orders of intelligence over the course of growth. The two (identification and structures of intelligence) are understood to be one in the same. Although this research is adequate to portray certain ideas, it is also partial. It is important to point out that this level of research enacts a truth that only holds solid from a perspective wherein identification is embedded in the personality planes. As identification deepens progressively into the soul and monadic domains (indigo altitude and beyond), one realizes that identification and intelligence are not the same thing, but can actually be clearly differentiated. This means that we need to find a deeper, more nuanced way of understanding the relationship between these two lines.

As discussed in Chapter 13, the perspective we take here understands each line of intelligence to be developed through the interaction of the UL consciousness, the UR body, the LR collective systems (of energy and matter) of a particular plane, and a LL shared culture. That is, it is the interaction of consciousness with the body, energy-systems and culture of a particular plane that helps to form the various lines of intelligence—cognitive, emotional, interpersonal, psychosexual, moral, etc.

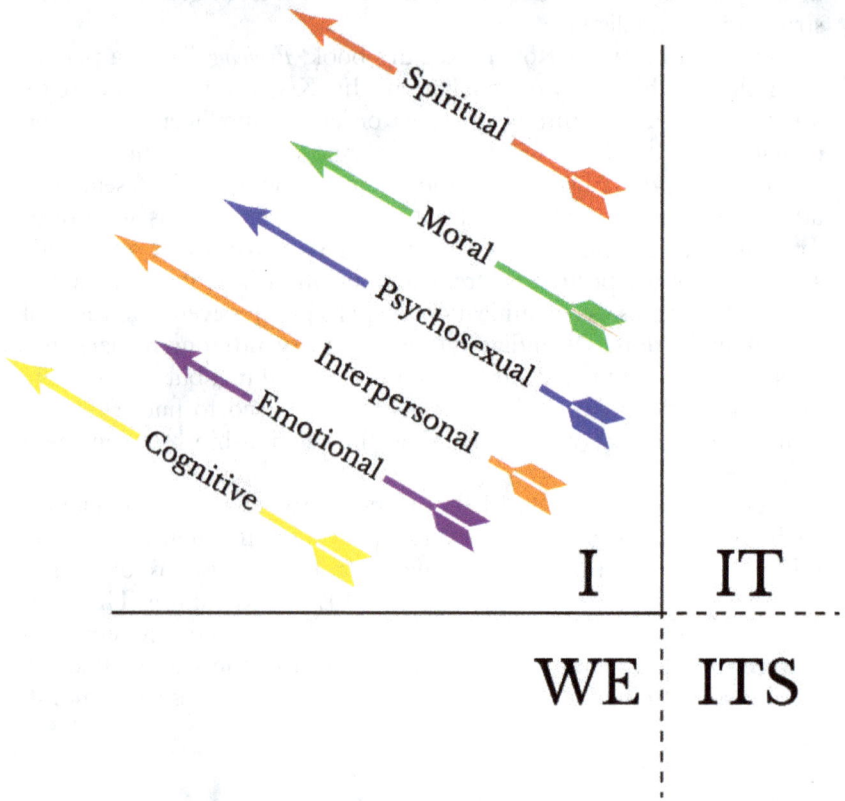

Figure 22. Graphic showing the multiple lines of intelligence.

From this graphic, and as has been presented in both Howard Gardner's theory of multiple intelligences and Ken Wilber's Integral Theory, we can see that we have multiple lines of development across the different domains of our being, and we can see that all are able to develop relatively independently as the consciousness aspect of our altitude of identification is focused through the planes.

This is the basis of those sometimes amazing and sometimes pathological examples we come into contact with in life, where people whose development across the different lines lacks balance, or *decalage*. Hypothetical stereotypes (though definitely with real world referents) are the genius whose cognitive line of development is through the roof, but whose emotional and interpersonal lines leave a lot to be desired; or the person whose emotional line of development appears to have extended into deep territory, but who consistently demonstrates an overly egocentric morality. Of course, taking us back to the definition of the self as both an integrative function and a line of development, we can see that it is the role of the self to maintain a nexus of integration across all of these different lines of unfoldment, but that crucially also, the self, in each of its depths, cannot be *reduced* to the lines of unfoldment.

When considering the relationship between identification and various intelligences, it is important to note that intelligences can be considered:

a) As possible of development *within* a particular plane that consciousness is operating on

b) As the extension of lines of intelligence *across* planes (i.e. from one plane to another)

c) As lines particular to a specific plane, but that develop into deeper expressions without changing plane as a result of a deepening of identification

So, for example, if we are thinking of the development of intelligences through (a) above, we would say that as identification shifts from one bodymind into a deeper one (personality to soul, for instance), there would then be the opportunity for the development of lines of intelligence particular to the soul planes (intuitive mind for the higher mental plane and intuitive mind for the buddhic plane, for instance). This example would suggest that a person's altitude of identification, whether that is, for instance, in the astral or mental faculties of the personality, or in the soul bodymind, *determines* the lines of intelligence they can develop. So, for instance, we would understand kinesthetic intelligence, such as might be expressed in dance or martial arts, to develop as the result of the focusing and operation of consciousness on the physical plane and in the physical body. Emotional intelligence would be the result of the focusing and operation of consciousness through the emotional body, though the mental body would be involved here too (emotional intelligence requires the capacity for 3rd person objectivity and rationality as well as sensitive feeling-centered intelligence). The intelligence structures considered often in Integral Theory, such as those ranging from red egocentrism to amber traditionalist to orange modern-rationalist to green post-modern pluralistic and turquoise integral, would then be understood to unfold as consciousness is focused on the mental plane (expressed in worldviews) and then increasingly across all the planes of personality in an integrated way. When we speak of such lines as Ken Wilber discusses in his consideration of the higher structures of intelligence included in the last chapter, we would see illumined mind, intuitive mind, overmind and supermind as developing through the focusing and operation of consciousness, now that of the soul and monadic bodyminds, on the soul and monadic planes (higher mental, buddhic, lower/higher atmic, and monadic respectively). This can be see in Figure 20.

If we were exploring the relationship in connection with (b), we could hypothesize that as identification shifts from one bodymind to another (personality to soul to monad) there may be lines of intelligence that extend all the way through, though changing their surface form of expression on the way. This is illustrated in the figure above. One example here would be the line of cognitive intelligence moving up from personality cognition into soul cognition and monadic levels of cognition. Again, this relates to Wilber's self-report of the higher structures of intelligence included in the last chapter. These levels of cognitive intelligence might look very different, but they are all the same line. Another example is Cook-Greuter's work on self-identity or ego-development. Other than cognitive development, her work is the only other line that begins to move vertically out of the mental plane and into the buddhic plane, in the graphic above. It seems that there may be some lines this is possible for and some it is not.

If we considered (c) above, and in relation to a shift in identification from personality into soul, we would say that that shift opens up the possibility of a deepening of the lines of emotional, cognitive, moral, psycho-sexual, and social intelligence, for instance, into forms rooted in the illumination, love, heart and wisdom of the soul. We would say, however, that in order for those different line structures to mature into fullness *they would need to be engaged and enacted*—just like muscles. This, plus all the examples above, again shows that shifts into deeper altitudes of identification open up the door for the development of new and deeper forms of intelligence, *but they do not take you through it.*

Additionally, it seems that the unfoldment of the intelligences of a particular altitude of our being (personality, soul, monad) seems to play a key role in whether a transition to a deeper bodymind must remain a temporary peak experience, or is able to be stabilized. What this means is that in order for the shift into a deeper level of identification to *stick*—to be stabilized—development across the lines of intelligence for that level of bodymind and plane (cognitive, emotional, moral, kinesthetic, for the personality, for instance) needs to have enough balance to be able to support it as a firm foundation.

To understand this, we can use the analogy of a wicker chair. A wicker chair is constructed of streams, or lines of plant stem fibers, and will only be able to support the weight of the one who sits on it (the soul, for example, if we are feeling into a transition in identification from personality to soul) if the plant stems that form its foundation (the various lines of intelligence-development at the personality altitude) extend deeply enough around and through the whole structure (the personality bodymind, in this instance).

If they do, the chair can provide a stable, strong, and comfortable foundation upon which the deeper altitude of our being, the soul, can be grounded. If they do not, then the chair is unstable, will tear and contain holes, and will not be able to support the weight of unconditional love-wisdom or power of the deeper levels. We feel that this kind of picture is often seen in spiritual teachers whose identification, in large part, genuinely abides in the eternal light and incandescent love of soul, but whose full abiding in that depth is prevented by under-development or shadow material in certain areas.

This wicker chair example also holds true for radical awakening. If the foundation of structural intelligence development is not strong and thorough enough, there will be no foundation stable enough to support the expression of the Absolute in an embodied form.

When the lines of intelligence development across the various plane-levels of a particular altitude of identification are sufficiently balanced, however, and the chair *can* support the weight, identification is able to be abstracted into the deeper level, the soul for instance. Then, the integrated personality becomes the foundation for embodied incarnation—healthy, integrated, balanced and whole. That is, as is the integral flavor of development itself, and when all runs smoothly and healthily, the former altitude of identification is *transcended, yet included.* This again governs the pattern of unfolding at the especially advanced stage of identification in which the transition is from soul to monad and beyond.[137]

There is an important historical point to make in this regard, one that honors the non-static evolutionary rooting of the structures of intelligence and makes sure not equate them with levels of identification. In taking the above position we are not suggesting that in order for a shift in identification from the personality to soul to be stabilized all the lines have *always* had to be developed into the highest structures available today in the collective quadrants, such as the post-modern or integral bands, for instance. That would not make sense, as according to Don Beck's research those structures were not developed until the latter half of the 20th century. Rather, we would suggest that throughout history, in order for a shift in identification to be stabilized, the various

---

137. There is a caveat to make on this point, which is that once the transcendence has occurred, the lower level of identification, particularly physical embodiment, can only be retained and included if its level of evolutionary complexity is sufficient to support and house the deeper bodymind. As we will go on to explore further later on, this has not always been the case, and thus the liberated beings who have gone before, such as those who now stand as the masters of the planetary meta-sangha, were not able to remain embodied whilst still transcending.

lines of a particular depth need only be unfolded into those structures that are the highest available *at that time*.

Radical Awakening and the Lines of Intelligence

There is a small link between radical awakening and the lines of intelligence. In general we often say that one can deconstruct reality at any point, no matter what their level of intelligence unfoldment might be. Now, this is true for the most part, but at least one caveat needs to be in place. Radical Awakening requires the stabilization and development of a healthy and stable ego. Healthy and stable ego development is the result of very early stages of structure-stage cultivation. This fact is precisely why Jack Englar came up with the phrase: "You have to have an ego before you can transcend it."

Just as a fundamental level of intelligence development is required for radical awakening, there is one inverse relationship between these that might also be true. It may be the case that the highest levels of intelligence development *require* certain degrees of radical awakening. Just as we saw how certain levels of identification are needed to open the door for the development of structures of intelligence, the same may be true with radical awakening. For instance, the supermind structure of intelligence (5th or 6th stage of Integral consciousness approaching Clear Light altitude) may require a realization of the deepest vantage point of Awakened Awareness. Similarly, there may be particular developments in radical awakening *required* for other advanced stages of structural intelligence development.

A second link, as we saw earlier, is the fact that no matter how awake a being might be, as long as he/she is still functioning through the mental plane as the main bandwidth of interaction, the structure of development of the particular person will influence the way that person *interprets* the experience. This is expressed in how the intelligences of the mental plane often show up as worldviews. The center of gravity of one's structure of development, as we have seen, determines the world that he or she enacts. In this way radical awakening is subject to interpretation.

This is an important point. It preserves the fact that radical awakening can occur through any structure of intelligence, but that that awakening will then be subject to being interpreted *through* that structure. Someone operating through an orange modern structure of intelligence who had had a genuine radical awakening experience would attempt to interpret it through a rational, scientifically-plausible lens, for instance, and might thus not be able to account for or hold the trans-rational aspects of the experience. Someone operating through an amber traditional structure, with a Christian background, who had an experience of relational contact with a subtle being through penetration onto the soul planes might interpret that experience as contact with Jesus or an angel, for instance. This is the Rainbow of Enactment described in Chapter 13.

Both of the above points necessarily relate also to how clean the personality, soul and monadic bodyminds are of any pathologies, areas of contraction, and shadow that may color them so as to cloud and obscure Awakened Awareness' transparency to itself. As some of the greatest teachers and realizers of the recent times have demonstrated, radical awakening is no cure for shadow. Rather, if shadow is present it just means that the vehicle through which the realization is expressed will be more or less colored by the narcissism, projection, neurosis, rule/role confusion or identity inflation, for instance, that might be related to that shadow material. This is why it is so important that we make the sincere commitment to engage not just with the process of *waking up*, but of *growing up*, *cleaning up* and really *showing up* for ourselves, and the whole planet, too.

Plane Access and Lines of Intelligence

In connection with the relationship between the development of structures of intelligence and plane access, here again the former can only be seen to enhance the latter. It is wise to note though that the development of structures of intelligence will partially determine the *content* enacted on any given plane. This is the point made above. There is also a relationship between these two when the development of intelligence structures is linked to shifts in identification, which, as we said, opens the door to plane access but doesn't take you through it.

Part 6. The Journey into the Kosmos

The endlessness of the path is so very beautiful!

(Master Morya)[138]

---

138. Roerich, 1930, stanza 357.

Earth is Eden

Chapter 17
The Initiations and Kosmic Paths
Jon

As humanity increasingly evolves, the channels for both trans-lineage dialogue and the release of awakening from being tied to any particular tradition are breaking open more and more. In this section, we suggest that the most leading edge and potent forms of spiritual living engage unfoldment across both the radical and evolutionary vectors, including all of the evolutionary lines simultaneously. This is the integration of the siddha and the shaman, or the mystic and the occultist. To explore this, we track the stages of these most advanced attainments, up to mastery and then out into the kosmos, as expressions of our potential across all vectors and lines actualized to profound levels of development.

The Path of Initiation

In the esoteric traditions generally, there is a vast array of stage models for understanding progress on the Path. Such stage models can relate to any of the vectors and sub-lines independently, or to a number of other continually developing indices. With all the diversity, attempting to locate any deep structure upon which they rest can often be difficult. As the advanced levels start to be unlocked, however, spiritual development becomes a function of both the radical and evolutionary vectors, and the sub-lines of the latter simultaneously.

Here the picture becomes clearer, and we can start to see that there is a convergence among many of the traditions on a set of distinct awakenings, shifts, developments and openings that beings start to experience sequentially. Examples of their representation in the traditions are the Four Stages of Enlightenment of Theravada

Buddhism; the Five Paths (Marga) and Ten Levels (Bhumis) of the Bodhisattvas in the Mahayana and Vajrayana Schools of Buddhism; the Sufi Stations ofbodyu the Soul; the Vedantic Seven Stages of Wisdom (Bhumikas) as discussed in the Varaha Upanishad; and the Trans-Himalayan teaching on the Path of Initiation. These Stages, Paths, Stations or Initiations, in their earliest forms, generally commence at the beginning of the Path, though they unfold into mastery and beyond.

The extent to which these unfoldments demonstrate transparent correlation across the traditions is an amazing testament to their existence as deep-level processes of cross-cultural significance, though necessarily expressed differently by different cultures in different places on surface levels. A tabulation depicting how they have been represented in three of the traditions along with some of their corresponding characteristics can be found below. In relation to this table, it should be pointed out that in these stages, you will notice some unfoldments which relate more to radical awakening than evolutionary becoming. This is owing to the fact that state-stage shifts in radical awakening are necessarily entailed in the advanced structure-stage shifts in evolutionary awakening, as was explored in Chapter 16.

Buddhist Five Paths	Sufi Stations of the Soul	Christian / Trans-Himalayan Initiations	Qualities and Characteristics
The path of accumulation (sambharamarga)	Station of the Heart	First initiation— Birth	Setting the foundations of practice, discipline, the first percolations of awakening into the mind, unwavering commitment, bodhichitta
The path of training or preparation or joining (prayogamarga)	Station of the Soul	Second initiation— Baptism	Aspiration, devotion, love, equanimity, heart, Bhakti and tantric/transformational yogas may be emphasized. The beginning of the realization of selflessness. Emotional / desirous equanimity
The path of seeing (darshanamarga) – first bhumi	Station of Divine Secrets	Third initiation— Transfiguration	Mental equanimity, profound concentration, power, clarity, knowledge, integration, end of physical karma, Jnana, Raja and Tantric Yogas applicable. Breaking through identification with the mental body—bringing deep intuitive opening. Direct nondual contemplation and path of self-surrender now accessible. Karma Yoga perfected. Tantric practices still relevant in meditative and daily practices.
The path of intense contemplation (bhavanamarga)	Station of Nearness to Allah	Fourth initiation— Crucifixion / Arhat	Trans-rational intuition, equanimity, profound (higher buddhic) love and wisdom, selflessness, stabilization of radical awakening, Rigpa (Dzogchen), Christ-consciousness, Sahaja Samadhi, end of emotional/astral karma, Jnana Yoga, Tantra, Agni Yoga. Breaking identification with the intuitive, spiritual self (soul or discriminating self)—leading to constant God-consciousness. Activity arises as wu-wei (Taoism), or 'choiceless action' spontaneously arising in harmony with Spirit.
The path of liberation or no more training (vimuktimarga)	Station of Union with Allah	Fifth initiation – Revelation/ Mastery	Self-realization, absorption in Trans-(Etheric) human Beingness, Nondual Consciousness, mastery, completion of human karma, experiential revelation of planetary Destiny, siddhas (perfected beings), Body of Light (Dzogchen). Exhaustion of mental karma, culminating the process of complete transcendence of personality/ego identification. If this level is fully integrated with the physical body, it results in the 'Body of Light' or 'Great Transfer', the transformation of the material elements into their inner essences.

Table 4. Table showing correspondences between the advanced stages of the path as described in terms of the Christian narrative, the Trans-Himalayan initiations, the Five Paths of Buddhism, the Sufi Stations of the Soul, and the characteristics of each of these stages.

Evolution moves both slowly and occasionally with great leaps, and these initiations can be understood as times when those appropriate circumstances have been met for such quickenings to be able to take place. Earlier, we described initiation as a ubiquitous law of the soul planes, and this is an important point to unpack. In the Trans-Himalayan teachings, these initiations are understood as soul plane ceremonial occurrences in which human beings, in groups and as individuals, receive dynamic transmissions of higher energies from beings dwelling on higher planes. This empowers their capacity to hold and transmit the current of Life. In such initiations, the initiate, whether individual or group, is charged with the pure voltage of higher realties. This serves to instantaneously integrate higher frequencies of being that have previously been the goal of attainment so that they then exist as the basis of operation for new stretches of the kosmic evolutionary path. In this way, each initiation marks a point at which what has previously been a peak experience is integrated as permanent trait.

As such, these initiations have both a *process* and an *event* aspect to them. The process aspect means that a particular initiation may occur very gradually over a number of lifetimes. The event aspect means that even with this being so, the gradual process will normally lead to a rapidly experienced shift at a particular point, perhaps immediately, or overnight, into a whole new experience of being. That point will mark both the consummation of an entire period of evolution, and the empowered entry into a new one. There are innumerable reports of these kinds of phenomena throughout the biographies of the great shamans, saints, sages and siddhas, and these process and event aspects could be seen to relate to the gradual and sudden path distinction also.[139]

These initiations are holons that arise across all Four Quadrants and both vectors. In relation to the Quadrants, each initiation is a point of stabilization at a new depth of interiority, which is able to be abidingly enacted in increasingly enlightened objective behavior via the activation of a subtler energy body and receptiveness to new, more powerful energies (personality, soul, and monadic). Additionally,

---

139. Aurobindo is one example here of having experienced a number of sudden and permanent shifts in his lifetime—most notably what he described as the "descent of the Overmind-Godhead" on November 24th, 1926. Krishnamurti's reported "life-changing" experiences, beginning in August and September, 1922, and continuing throughout his life, are another good example, involving profound and sudden shifts in his awakening, psychological transformation and the raising of vibration in his subtle energy bodies. Eckhart Tolle's suddenly experienced awakening and transformation in 1977 is another.

it is a new phase of shamanic entry and opening into collective modes of organization and community, on the personality, soul, and monadic planes, that are ever more expressive of that divine ethic. And each initiation allows new participation in a collective worldview and set of ethical injunctions that are progressively saturated with the radiant love and dynamic power of the soul-level and monadic realities.

In connection with the initiations' impact across the vectors, Djwhal Khul and Bruce Lyon describe the first two initiations as involving shifts along each of the sub-lines of the evolutionary vector,[140] whilst the 3rd initiation and onwards incorporates radical awakening also. The first initiation marks the commencement of the path for an individual. The light of their own soul has touched their mind enough for them to start to recognize the kosmos as intrinsically sacred and alive. They set their feet upon the way of conscious growth and development for the first time in a committed way, and generally that commitment expresses itself in a focus on physical disciplines and practices. Examples might be physical yoga, conscious dance, bodywork, or a focus on working with entheogens, such as ayahuasca, DMT, LSD, psilocybin mushrooms, or others.

The keynote of the second initiation is emotional purification. Now the individual has committed themselves to walking the path, they go through a process of purifying, healing and integrating their emotional impulses, desires and wounds. The soul pours the energy of pure love into the heart center of the individual and they come to see clearly all of those ways in which they contract personally and are unable to maintain the conduct of love, in relation to themselves, others and the world. Passing through the second initiation involves a profound transition from a life lived under the impulse of personal desire to one that flows from the heart and love. The initiate more and more learns to live from love. They find their group of fellow souls who are their family at the level of soul, and they start to unfold their unique path of service to the world.

Though openings to radical awakening may have happened previously, it is only at the third initiation that it starts to play an abiding role. This is not to say it cannot happen before this point. It absolutely can. But the third initiation is the point at which Nondual awakening to the naked Reality of Awakened Awareness takes on a more effortless and natural flavor, and begins to have effects on the flow and movement within the evolutionary sub-lines too. This correlates with Buddhist teaching on the Bhumis, or most advanced and expansive processes beginning at the Third Path, the Path of Seeing, or what Jesus is understood to exemplified in the Christian teaching at

---

140. Lyon, 2005.

the Transfiguration. As elucidated by the Vajarayana and Dzogchen teacher, Tsoknyi Rinpoche:

> Determination makes all the difference, because our rigpa [radical awakening] does fluctuate, experientially, until we reach the first bodhisattva level. From that point onward there is no falling back or straying into confusion. The path is now a steady, smooth journey. Before that, we forget, then we remember, we forget, then we remember— flickering back and forth. Don't expect that because yesterday we were introduced to rigpa, we recognized it as a glimpse, now we are set. A glimpse of experience isn't going to transform everything or carry us all the way to enlightenment. It doesn't happen like that. Only at the first bodhisattva level do you have true confidence, a confidence that does not waver any longer.[141]

Also from Tsoknyi Rinpoche on the relation between radical awakening and the Buddhist paths and bhumis:

> There comes a point when we are almost self-reliant, known as the path of joining, which is after the path of accumulation. When you arrive at the path of seeing, which corresponds to the first bodhisattva level, you are completely independent. Your own experience is then the actuality of rigpa, with no confusion any longer. From then on you are independent, absolutely self-reliant in proceeding onward to a buddha's true and complete enlightenment.[142]

According to the Trans-Himalayan understanding, at this third initiation (the Buddhist path of Seeing and first bodhisattva level), as well as the stable access to radical awakening that opens up, the personality is truly integrated for the first time, which allows it to become an open vessel for the soul to incarnate through. As this occurs, there is also the opening up of conscious relationship with the initiates and masters of Hierarchy. Shamanically, it stabilizes one's volitional plane access into the energy matrix of the soul planes. From an Integral perspective, this would entail one's intelligence structures moving into 3rd tier levels, such as illumined mind and intuitive mind—the intelligences of the soul. Additionally, the third initiation sees the individual rise to the heights of their world service. They contribute to the world some project,

---

141. Tsoknyi Rinpoche, 2003, p. 204.

142. Ibid, p. 205.

organization, movement or institution that embodies their realization of the evolutionary plan for all beings. In the Buddhist tradition this is the making of an Arya-Bodhisattva, and as said, the beginning of the extraordinary path of the bhumis.

While the third initiation involves stable access to radical awakening, so the initiate no longer has any doubt about the true nature of reality as Awakened Awareness, the fourth initiation sees that realization stabilize so fully that it literally burns through the last remaining illusions of separation—the soul. The fourth initiation sees the individual consumed totally in the realization of Nondual Source, and they start to become powerful transmitters of that realization to their surrounding environment. They let go of operating in the driving seat of the project, initiative or movement they contributed to humanity when passing through the third initiation, and yet they become even more powerful inspires of it through their transmission of Source-energy into it as they simply live an awakened life. Momentously, at the fourth initiation they transition out of the planetary culture of humanity, and into the planetary culture of Hierarchy, and stabilize their working through 3rd tier levels of soul intelligence whilst having full access to the monadic, atmic, buddhic and higher mental planes. At the fourth initiation the initiate's consciousness is now focused on penetrating their the last remaining layers to be integrated at the emotional and physical levels with the radiance of their awakening.

This is completed at the fifth initiation. Then, the full stabilization of radical awakening to Absolute Presence is augmented by the abiding abstraction of identification into the power and planetary Purpose transmission of the monad. The initiate's evolutionary awakening penetrates into the sacred Purpose of the Earth as a whole as it is envisioned in the heart-mind of our planetary Logos, and they become an empowered agent of that Purpose. At this point the entirety of initiate's embodiment has been irradiated by the light of realization. The awakening that was established at the third initiation has now matured into full enlightenment with its penetration all the way through the mental, emotional and physical layers, leaving nothing on those planes operating out of harmony with Reality. This initiation marks the beginning of conscious relationship with the buddhas composing Shamballa; the stabilized shamanic capacity for vision into the energy matrix of the monadic and logoic planes; and the unfolding of experientially kosmocentric intelligences. At this point, all planetary form, subtle, and gross, has been unclothed as Eden—total Goddess. This is understood as the achievement of mastery in the Trans-Himalayan teachings. For the esoteric Christians it is the Revelation; for the Sufis it marks the station of Union with Allah and in the Buddhist tradition it is the Path of Liberation, or No More Learning.

Beyond the Kosmic Physical Plane: The Kosmic Paths

According to the Trans-Himalayan teachings, the path of initiation does not stop there. In fact, as a centerpiece of the Trans-Himalayan teachings and as has been a teaching held by some of the traditions, astonishingly, it *never* stops unfolding. Within the *kosmic* community and scale of awakening, terms like 'mastery' are relative, and whilst the fifth initiation might seem like the consummation of the Way from one perspective, from another, it is just the beginning. From a kosmocentric perspective, Earth mastery is the *first* initiation.

As traditions such as the Buddhist, in the form of the *bhumis*, explore, as well as the Vedantic, Trans-Himalayan, and the Ancient Egyptian, when radical awakening to the deepest vantage point has revealed all of the kosmic physical plane as Nondual Perfection; when shifts in identification have been surrendered into so fully that one's abiding identity resides as the power and Purpose-transmission of the monadic bodymind; when relationship has been stabilized with all the multidimensional communities of the kosmic physical plane as Self-reflections of the One Life; when one's shamanic penetration and opening to penetration has encapsulated the entire kosmic physical plane; and when one's siddhis of intelligence have opened into the kosmocentric, not conceptually but experientially—so that one can literally *feel* the Life-pulses of the kosmos in the nerve-fibers of one's bodies, there open up whole new levels of exploration, identification and development, out into the Nondual kosmos. In the words of the Vedantic sage, Krishnananda:

> We have attained to a unity by bringing together all particulars into the universal. Now we transcend even the universal physical for the sake of the attainment of the universal psychic or the astral; transcend that also, later, and then reach the universal causal [high atmic/low monadic]...[143][144]

In coming to remold our own personal cosmologies so as to hold space to receive the profundity of this vision, we need to remember that *the planes extend indefinitely*, with progressively subtler planes of energy-matter transcending, including and interpenetrating the denser planes. As plane access pierces the monadic altitudes (lower atmic, monadic proper and logoic), the entire spectrum of all seven planes is

---

143. Krishnananda, 2008, p. 39.

144. Here Krishnananda hasn't differentiated the evolutionary and radical awakening vectors—hence his use of the term "causal".

revealed as the internal vibratory states of the kosmic physical plane—the incarnated universal Mother; visionary penetration into planes of greater subtlety, expansiveness and inclusiveness of consciousness, energy, relation, and community throughout the kosmos occurs.

All of this involves an incredible shift in experience, especially along the evolutionary line of plane access, where a new scale of planes reveals itself beyond the kosmic physical plane. This development consequently calls forth profound shifts in relation to radical awakening, and to the other three evolutionary awakening sub-lines (identification, relationship, and intelligence cultivation).

To begin with radical awakening, while this is awakening to Reality *beyond* the entire spectrum of time and space in the manifest universe (on all levels), that awakening still happens *within* relative time and space. That is, it occurs within some particular plane of vibrational energy-matter and at some point in the history of the evolutionary process. The plane on which it occurs is the kosmic physical plane (on any of its seven sub-frequencies), and so the extent of the manifest universe that is able to be revealed as the Nondual Great Perfection *extends only to the kosmic physical plane*. In the words of Djwhal Khul with Alice Bailey:

> When the student realizes that the great universal Oneness which he associates with monadic consciousness, is only the registration of impressions localized (and therefore limited) and defined within the etheric levels of the cosmic physical plane, he can perhaps grasp the implications of the wonder which will be revealed to the initiate who can transcend the entire cosmic physical plane (our seven planes of the human, superhuman and the divine worlds) and function upon another cosmic level. This is what the treading of the Way of the Higher Evolution enables a Master eventually to do.[145]

Thus, as the periphery of the kosmic physical plane is pierced, there unfold more and more kosmic energy-fields through which the master is able to extend their radical realization of the Nondual Self-Perfection of Reality.

In relation to the evolutionary awakening sub-line of identification, the consideration of deeper levels to realize than the monadic bodymind forces us to a) really begin to engage the top-down perspective on the reality of identification, rather than the bottom up perspective we usually assume, and b) recall the distinction mentioned in Chapter 10 between the monadic essence and the monadic bodymind. As we explored there, ultimately, the ever-present root of our identification-

---

145. Bailey, 1960, p. 363-364.

line is Absolute Source, which in its first explosion of radiance in the creation of the universe in relative time broke forth as Ishvara,[146] the One Universal Monad. This Universal Monad arises through the willed process of emanative Self-reflection, Self-objectification, and consequent though relative Self-division, which allows Reality to 'ray-forth' as the kosmos. The One Universal Monad emanates the galactic monads that incarnate through entire galaxies, which themselves emanate super-constellational monads, which incarnate through collections of constellations, which themselves emanate constellational, solar, planetary and intra-planetary monads, which all incarnate through their respective fields until *and beyond* that 'point' at which we have what we understand as a 'human monad'. In the words of the Trans-Himalayan teacher and scholar, Michael Robbins:

> Any given M/monad [capitalization refers to a great monad, such as a planetary, solar or galactic monad; non-capitalised refers to a human monad] is a specific Self-perception of a certain magnitude. The Self Who is perceiving is the One [Universal] Monad. (For instance: the greatest Galactic 'Gods' are Self-perceptions of very great magnitude. A Planetary God is a lesser magnitude of Self-perception. A human being is a Self-perception of relatively slight magnitude). All these 'G/gods' are Self-perceptions of the One Monad. None of these 'G/gods' is Self-perception entire.[147]

The evolutionary Path of Return can thus be understood as the path along which these 'lines' of emanation are retraced and the monadic essence progressively recognizes that it is and has always been the One Reality, emanating as an infinity of kosmic Self-reflections on greater and greater scales. Consequently, as the eye of the monadic bodymind is opened fully enough to see beyond itself, there is the realization that the abstraction of its identification back into its deeper roots within the manifest kosmos extends far beyond the kosmic physical plane. Indeed, at this stage, the kosmic physical plane is revealed as the *womb* of the kosmos, birthing the Self-Realized Buddha that is Reality itself into a galactic field of maturation, growth, and service.

---

146. For us, Ishvara or Vairochana represents the monadic identification at the highest octave evolutionarily awakening (universal octave) and the deepest vantage point of individuated awareness in the radical awakening vector.

147. Robbins, 2004, p. 7-8.

What is Integral Enlightenment in an Infinitely Unfolding Kosmos?

According to the conventional Integral model, Ken Wilber has suggested that a post-metaphysical definition of enlightenment would define it as union with all states and structures available at a particular point in history (the states relate to the States of the Absolute—Nondual, Causal, Subtle, Gross—that are accessed in fullness through radical awakening and stay static through time, whereas the structures develop and unfold over the course of evolution). If we apply that to the Trans-Himalayan teachings and the model we are using here, that gets a lot harder. It's not that we disagree. We don't. It's rather that the definition needs quite a bit more nuance.

The reason for this is that we are using a multidimensional model that incorporates a) a top-down perspective on identification that recognizes more inclusive spheres of identity already operating at the highest levels of the universe and into which our own identity is progressively retraced and synthesized along the line of its emanation; b) a multidimensional understanding of the communities of lives already residing on and kosmically serving from the various planes of the kosmos; which, c) have no 'upper' limit; and d) a recognition that the structures of intelligence conventionally used in Integral (up to turquoise) do not go higher than the mental plane, and that there are both further intelligence structures on higher planes and beings abiding on those planes who have already unfolded them.[148] In brief, the Trans-Himalayan teachings posit beings operating through evolutionary structures way beyond what humanity can conceive of at the current time, with no limit prescribed for where they end. In that context, how do we define a final and completed enlightenment?

When exploring a two-vector Integral Trans-Himalayan definition of enlightenment, one option is to define it as stabilized radical awakening plus evolutionary awakening up to and including a specific marker of evolutionary developmental attainment. This might be the additional unfoldment of all evolutionary lines up to the present height of *human* evolution. This would have reference to the 4th initiation, as it is then that the being is understood to transition from humanity into Hierarchy.

---

148. We honour Integral's evolutionary perspective. We insert the caveat here though that such beings would likely not have developed the most advanced mental plane intelligence structures available to humanity today, as their ascent onto higher planes within the planetary aura (e.g. the planetary meta-sangha of Hierarchy or Shamballa, for instance) would have taken place before the development of those structures.

Another possibility would be to use the marker of the levels of evolutionary unfoldment presently related to the attainment of mastery at the 5th initiation, or the mastery of and freedom from the entire physical plane at the 7th.

These possibilities include Wilber's important element of contextualization as to how we define enlightenment, a move that is important owing to its capacity to incorporate the evolutionary perspective. As Wilber has noted, enlightenment can no longer be statically defined purely in terms of stabilized Nondual realization along the radical awakening vector, but must also include the degree of structural development, or what we describe as evolutionary awakening embodied by a being. For us, that means that yes, radical awakening does resolve every depth of reality available to consciousness into the Nondual Great Perfection, but how far (i.e. onto how many kosmic planes) that realization extends *is a function of one's depth of identification, relationship cultivation, and plane access.* From this we begin to glimpse the sacred recognition that while radical awakening remains as that utterly precious opportunity available to every human being for unmediated awakening to the Absolute, the evolutionary path of what happens next is literally *infinite.* That path extends into numberless planes of kosmic unfoldment and service in which communities of awakening throughout the universe are also engaging with this incredible adventure. This is the path of infinite kosmic reanimation and reclamation of the essential Unified Source, and the perpetual discovery and creation of novelty.

When frameworks do not include this multidimensional evolutionary perspective on the kosmos, we tend to end up with those presently pervasive narcissistic narratives which present the human beings deemed to be the most structurally intelligent and spiritually interested as the 'leading edge' of kosmic evolution. It is true that the path is not pre-given, but forged and sculpted through evolutionary unfoldment so as to become kosmic habits—freeways of developmental journeying amid sprawls of side roads that most often lead to dead ends. Through the eye of the kosmic-scale model of radical awakening, identification, relationship, plane access and intelligence unfoldment we use here, the ballpark in which this whole wonderful process is unfolding increases by a monumental order of magnitude. We see that the Path extends into regions of Nondual evolutionary unfoldment far beyond our current sight, and magnificently there have always been beings to walk it.

These highly advanced paths may as yet remain analogous to single lanes marked out with traffic cones, rather than fully-fledged freeways. This is owing to what seems to be the comparatively small number of beings of sufficient development to unfold them beyond that which we currently see as representative of our highest potential. In our exposition of the kosmic paths below, this point is expressed in the Master Djwhal

Khul's teaching that the Paths along which masters may either continue their trail of unfolding, within the aura of the Earth or out into the galaxy, are presently seven in number. However, until recently only two paths had been engaged sufficiently to have been laid down as deep grooves in the kosmic riverbed of unfolding. Having noted that now seven paths stand available to the liberated master, Djwhal Khul additionally states that increasingly, with growing numbers of liberated men and women blossoming into their own mastery, two more paths are being etched out in the fabric of the kosmos, making a total of nine.[149]

From an Integral perspective, it is only as more and more beings grow through and engage with the higher altitudes available on the kosmic paths that their UL internal experience and UR modes of individual behavior will be translated into LL collective cultures and LR embodied systems of interrelation. As of yet, their lower collective Quadrants, in their color, tone, fullness and interior composition, remain nebulous. However, with the number of human beings who are moving into the advanced initiations of awakening, identification, relationship, plane access and intelligence unfoldment at this time, humanity may increasingly ground, expand, elaborate and furnish these kosmocentric structures of development. Should we then together make it through the crises that currently loom so large, we may see the emergence, more and more, of humanity as a kosmocentric culture and civilization in the universe.

The Kosmic Paths

Radical awakening reveals all that arises in time and space as transparent expressions of Source. Upon this realization and its stabilization, relative reality, arising as all time and space, does not disappear. Time and space continue to arise but are now seen clearly for what they are: the Great Perfection.

In this section we are daringly speaking of that which occurs *after* liberation. When the Buddha was questioned on what occurs subsequent to nirvana he chose to remain ambiguous, only replying: "This teaching is profound, difficult to see, difficult to understand, calm, excellent, beyond the domain of reason, intelligible to the wise".[150] In the Ancient Egyptian tradition, evidence for these paths can be found in the teaching concerning the opportunities that opened up for the

---

149. Bailey, 1960, p. 412.

150. Chau, 1999, p. 127.

initiate once they had stabilized their shifts of identification into the *Akh*—the monadic essence. For the Ancient Egyptians, upon liberation the *Akh* was understood to be able to journey back into the kosmic heavens from which it came and resume its status as a star. Indeed, in this we might find a deeper understanding as to why the sky shafts of the Great Pyramid were aligned with the in-pouring rays of particular constellations, why the ancient initiatory Pyramid Texts themselves were known as "the akhifiers",[151] and why on the ceilings of so many Ancient Egyptian temples we find only the engravings of stars.

We might imagine that in the ancient esoteric traditions of the past, the number of beings ready to tread these paths was very low. Now, however, on account of humanity's evolutionary process and the increasingly manifold number of human beings on the planet who are genuinely penetrating into these levels of abiding realization, this is changing. More masters, to whose awareness this extraordinary reality of the unending nature of the Path has dawned, have begun to share on it. Again, Djwhal Khul:

> These seven Paths, when trodden, prepare a man to pass certain cosmic initiations, including those upon the Sun Sirius…. Each of these Paths eventually leads to one or other of the six constellations which (with ours) form the seven centres in the body of the ONE ABOUT WHOM NAUGHT MAY BE SAID [the Galactic Logos incarnating through our entire Milky Way galaxy]. Those adepts therefore who stay for a prescribed length of time upon our planet are a correspondence to those greater initiates who remain for many kalpas within the solar system, taking certain mysterious initiations concerned entirely with solar evolution. Their work is concerned with the system as a centre in the body of that Existence Who vitalises the Logos of our own system.[152]

And elsewhere, with Bruce Lyon:

> It is the role of initiates past a certain degree to maintain the cosmic connection between civilisations in the galaxy and increasingly, between galaxies.[153]

As Djwhal Khul says in the first of the above quotations, these Paths are seven, and he points to their nature as kosmic shamanic

---

151. Allen, 2005, p. 425.

152. Bailey, 1925, p. 1242.

153. Lyon, 2010, p. 78.

trails of journeying by describing how three lead to penetration of the kosmic astral plane, three to penetration onto the kosmic mental plane and one to penetration onto the kosmic buddhic plane. Each of them ultimately involves work with *form* on kosmic levels. Prior to liberation, it is interesting to note that so often it is the formless which is emphasized, and yet once a being is fully liberated—once Awakened Awareness, arising as a master, has awakened to its root that transcends, includes and arises as the entire relative reality moment-to-moment the orientation reverses. It then turns back with an indomitable love, will and commitment towards the universal spectrum of form. This is the making of a kosmic bodhisattva.

Of the seven paths he has described in his work with Alice Bailey and Bruce Lyon, six lead to other centers of service within the galaxy, while one involves the master choosing to stay within the aura of the Earth and the kosmic physical plane so as to serve the process of evolution and awakening in all kingdoms. This can be understood as a kosmic correspondence to the point made in Chapter 10 that in the Trans-Himalayan super-physics of evolutionary development, once a being has mastered 2/3 of a plane so that consciousness directs energy-matter on that plane, they may choose to transition to the next, more subtle plane, or stay to master the last third. Masters of the 5th initiation can be understood as beings who have mastered 2/3 of the entire kosmic physical plane; thus at the 6th initiation, which is known in the Trans-Himalayan teachings as "The Decision", they choose whether to transition onto more subtle kosmic planes of service, or to stay.

Of these beings that choose to stay, their continued unfoldment entails their abstraction of identity into the ocean of love-energy of the kosmic astral plane, but their retained abiding within the aura of the Earth on the kosmic physical plane. They enter into the collective culture of Shamballa, and yet work within Hierarchy for the evolution and awakening of all life-spheres on Earth. From their already-established Nondual awakening to the Great Perfection of the whole of the kosmic physical plane, their abstraction of identification deeper into the kosmic astral allows the development of kosmocentric intelligences, expressed in service *within* the kosmic physical plane. Their work is to develop sensitivity to extra-planetary relationships, other galactic civilizations, as well as to those Logoi that have chosen to ensoul planets, solar systems, constellations, and galaxies. As their abstraction of identity into wider and wider spheres of kosmic life continues, they expand their mountain-like identification with, and electric transmission of, the kosmic planetary Purpose of the Earth to that of our solar system and eventually our galaxy, all *without transcending* the kosmic physical plane. This is understood to be a profoundly challenging path of deep sacrifice, and is called by Djwhal Khul "The Path of Earth Service". In this process these masters choose which sphere of evolutionary complexity (quarks to atoms to molecules to cells to organisms, plants, fauna, humanity…) to work with on Earth so as to facilitate the emergence of the particular divine expression that the planetary Logos incarnating through the Earth is seeking to express through that kingdom.

Those masters who choose the second path abstract both their identity and plane access onto the kosmic mental plane. From there, they are able to embrace in Nondual glory the kosmic mental, astral and physical planes as unending Awakened Awareness. They open up relationships with vastly realized communities of lives on those planes and engage the development of kosmocentric intelligences to what must be an astonishing degree. These masters are understood to work with the electro-magnetic polarities of the kosmos on extremely subtle levels of energy-matter, eventually penetrating onto the kosmic mental plane. They are described as coming to eventually relate the enlightened energies transmitted from and between the Pleiades and the Great Bear to our solar system and the Earth. Both of these constellations are understood to be both ensouled by profoundly advanced beings, and to be home to civilizations far in advance of our own.

The masters who choose the third path again learn to abstract their identification so as to penetrate into and incorporate into their Nondual awakening the kosmic mental plane. From this summit of abiding, and out of great compassion, they work within Shamballa, learning to transmit planetary Will and Purpose as it is envisioned within the heart-mind of the planetary Logos of the Earth to all lives and kingdoms. Simultaneously, owing to training they undergo within Shamballa, they develop the extraordinary siddhi of learning to envelop and ensoul an entire planet with their consciousness and energy so as to hold it as a field for the evolution and awakening of all beings therein. The path of training for these masters is understood to be extremely long, and to involve an eventual transition into the planetary bodymind of the planetary Logos incarnating through Venus. Over its course, they cultivate relationship with those planetary Logoi who at the beginning of this cycle of kosmic manifestation chose to ensoul each of the planets of our solar system. In developing relationship such as this, they gain experience in the methods used by each of those beings in projecting their Will and the envisioned Purpose for that planet into its field.

The fourth path is known as "the Path to Sirius", and the masters who engage this path abstract their identity and plane access onto the kosmic astral plane and anchor their presence within the energy-body of the giant sun, Sirius. Sirius has figured prominently in the ancient mythologies of a number of the traditions. It played a key role in the Ancient Chinese, Persian, and Babylonian esotericism. For the Ancient Greeks, the apparent movement of Sirius through the heavens was celebrated during the Eleusinian initiations. The Ancient Egyptians considered Sirius the most important star in the heavens and based their entire calendar around it. The Dogon Tribe in Mali have been shown to possess an extraordinary level of understanding and insight into the Sirian constellation. In many esoteric lineages, it is understood as the original home of humanity and the source of the Mystery Traditions themselves.[154] In the Trans-Himalayan teachings, Sirius is understood as a great nexus of outstandingly enlightened kosmic community, and as ensouled by an exceptionally enlightened kosmic Logos. In a manner that echoes the Sirian connection with our solar system and planet described in the traditions, Djwhal Khul states that the decisions and methods of the Hierarchy of liberated masters who remain on Earth to serve and guide the evolution and awakening of life here is itself telepathically guided by the community of buddhas on Sirius. The masters who choose this fourth path pass to this community where, from radical awakening on a colossal scale, they continue to move through kosmic initiations of unfolding.

The masters who take the fifth path abstract their locus of identity and penetration onto the kosmic mental plane. They thereby include all within the kosmic mental, astral and physical in the Nondual embrace of the Absolute, opening up relationship with communities of lives on those planes, and also unfolding kosmocentric intelligences. Their work is understood to relate to the prismatic shakti-transmission of the seven

---

154. The Dogon Tribe in Mali are a people whose oral traditions suggest they migrated to their current settling place from a number of different regions of Africa. They are reasonably well known globally owing to their possession of advanced astronomical knowledge concerning Sirius. Specifically, Western researchers spending time with the Dogon during the 1930s found them to consider Sirius to actually be a dual star system, with a much smaller second star orbiting the much larger sun every 50 years. Astonishingly, this replicates perfectly the scientific understanding of the Sirius system, but what has caused so much controversy is that the existence of Sirius B, the white dwarf that circles the much larger Sirius A, is not visible to the naked eye, had only begun to be mathematically inferred to exist in 1844, and was not photographed until 1970. To many, it remains a mystery how the Dogon could have been in possession of knowledge that only came to be known within Western astronomy from observations through high-powered telescopes in addition to advanced mathematics. We would suggest firstly that their possession of this knowledge not only supports the notion that humanity's understanding of the kosmos does not simply grow incrementally over time, but may have peaks and troughs to it. Secondly, we suggest that according to a 4-vector analysis, this knowledge may have been a product of subtle realm penetration and perhaps even subtle realm communication with beings who were able to pass on this knowledge to the Dogon, or perhaps to the Ancient Egyptians, who many, including some of the Dogon themselves, believe to be their ancestors. Interestingly, clear similarities can be found in their cosmologies and creation stories.

rays and to the perfection of their transmission capacity of the seven rays to and through the seven chakras of the planetary Logos of the Earth. It is stated that they then abstract their presence into the subtle energy-body of the sun so as to expand their transmission capacity to a solar scale, before exiting our solar system to develop the capacity to transmit to and through an entire constellation.

The sixth path involves an expansion of the third, with these masters abstracting their identification and plane access onto the kosmic buddhic plane so as learn to project their consciousness and energy to ensoul and envelop an entire solar system. They are understood to transition to the correspondence to Shamballa within the bodyminds of the Logoi presently ensouling the other planets of our solar system. They then enter the bodymind of the solar Logos who ensouls our solar system itself, before passing into the correspondence to Shamballa within the bodyminds of solar Logoi who are presently ensouling solar systems elsewhere in the galaxy. The masters who take this path necessarily learn to project their energy-bodies and consciousness extensively, so as to hold space for the Logoi ensouling planets to reside *within their being* as their embodied chakra system. The being we described earlier and who is known in the Trans-Himalayan teaching as Sanat Kumara is understood to be treading this path.

The seventh and final path Djwhal Khul discusses as presently available to liberated masters involves the abstraction of their identity and plane access onto the kosmic mental plane. From there, in their radically awake and kosmically experiential siddhis of intelligence, they direct the forces of karma for the entire solar system and keep open the lines of relationship in energy and consciousness with the ensouled constellation and the civilizations of the Great Bear, and the supermassive black hole at the galactic center. It is understood to be the destiny of these masters to work within the presence of the galactic Logos that envelops the entire galaxy with its consciousness and energy.

The nature of the advanced stages of development we are discussing here are profound. In any exploration of this sort, is crucial to remember that even though we are speaking of these paths at times in terms of choices made by masters and buddhas, the Reality that is their basis is very simple. Ultimately, what is unfolding here is just the increasingly unrestricted Self-expression and Self-display of the Absolute Conscious Light to itself through the entire kosmic domain. As we contemplate the vastness of the kosmic process we are a part of, may we do so from recognition of our True Nature as the Ultimate Clear Light that is the radiantly awake basis of every layer of the multidimensional kosmos, no matter how exalted.

Chapter 18
A Holonic Kosmic Ecology
Jon

Imagine for a moment what it would be like to step outside of the anthropocentric models of the kosmos that blinker our eyes so often in the current times. What if there were a cosmology that included all kingdoms of life and scales of the universe? The Integral Trans-Himalayan 'holonic kosmic ecology' explored here begins to open a world beyond the human and into the kosmic. In its presentation, we look to initiate a post-metaphysical approach to understanding some of the most esoteric teachings of the Trans-Himalayan canon. We hope that by doing so, we might build a bridge for these teachings to contribute to the kosmocentricity of an emerging universal spirituality (those interested in moving from a worldcentric to kosmocentric perspective might want to pay careful attention as this unfolds).

From the dynamic energy of Primordial Awareness arises an infinite holarchy of interiors and exteriors, singularities and collectives. The exteriors arise as multidimensional buddha fields, beginning with the sacred mandala of the universe itself, within which arise galactic fields, then stellar fields, then planetary fields, where on some incredibly special planet(s), there are the suitable conditions for the evolution of biological life. According to the Trans-Himalayan teachings and our previous discussions on the kosmic paths, on every level of the manifested universe each UR body has its incarnating UL consciousness, just as for every human being the personality, soul and/or monadic energy-bodies support the interiorities of personality, soul, and monadic consciousness. Thus, in the UL quadrant corresponding to each scale of UR body—human, natural kingdom, planetary, stellar, galactic, universal—there exists at each octave a single interior *consciousness.*

These consciousnesses are the hierarchically emanated Self-reflections of Absolute Source—Awakened Awareness. As we saw at the beginning of Chapter 10, Awakened Awareness's most primal radiation of effulgent Light in its first act of self-objectification arose as the One Universal Monad. This One Universal Monad then emanates as galactic monads, which incarnate through the hundreds of billions of galaxies existing in the universe. These galactic monads emanate super-constellational monads, which incarnate through systems of constellations. These super-constellational monads emanate constellational monads, which incarnate through constellations. These emanate stellar monads, which incarnate through suns and solar systems, and these stellar monads emanate planetary monads, which incarnate through planets. These planetary monads then emanate all the monads who unfold and evolve through the various life-spheres of evolutionary complexity on their respective globes.

All of this leads us to recognize seven different *octaves* at which kosmic Logoi are operating through various scales of emanated identification, with each of these potentially having personality, soul and monadic sub-levels of bodymind expression:

Octave	Incarnating Entity
1. Universal	Universal monad
2. Galactic	Galactic monad
3. Super-constellational	Super-constellational monad
4 Constellational	Constellational monad
5. Solar system	Solar monad
6. Planetary	Planetary monad
7. Human	Human monad

Table 5. Table showing the seven octaves, or scales of incarnation, through which kosmic Logoi, each with various degrees of evolutionary development, are taking form.

Each more materially complex octave of monadic emanation (from top to bottom in the above chart) can be understood to allow and involve an extension to the evolutionary complexity demonstrated and embodied in consciousness and matter. This stems all the way from the Big Bang, after which the cooling of the rapidly expanding universal space precitipitated the emergence of the first subatomic particles such as protons, neutrons and electrons, the former of which combined quickly to form the first atomic nuclei; then the combination of electrons with them and the consequent production of the first elements—hydrogen and helium—which allowed the arising and formation of the first structures of the universe: black holes and the dark halos of early galaxies. Within these there were then major bursts of star formation. From the cycles of star birth and death there came the sufficient diversification of elements for the formation of planets, such as Earth, where the suitable conditions for the emergence and evolution of biological life exist.

Normally in Integral Theory, and as is expressed in a Four Quadrant matrix, complexity of matter is correlated with complexity of consciousness. Therefore, in the LR of a conventional Four Quadrant matrix, we can see that kosmic bodies such as galaxies and suns, for instance, are understood to be *less* evolved than all biological life on Earth. This is interesting, as in the Trans-Himalayan teachings the beings or consciousnesses incarnating through galaxies and solar systems are understood to be vastly *more* evolved than any life on Earth.

This points to a paradox, whose reconciliation is found in an integration of Integral's holonic perspective, where all components of relative reality are understood to be wholes that are yet parts of greater, more inclusive wholes, and the Trans-Himalayan teaching's understanding of the process of universal monadic emanation. When this integration is in place, we can see holonically that Earth is a part of the solar system, which is a part of the galaxy, which is a part of the universe—all of these the evolving expressions of the Absolute. Simultaneously, according to a Trans-Himalayan understanding of monadic emanation, each octave of increasing evolutionary complexity embodies an extension of Ultimate Source's capacity for emanated Self-reflection to a new level. From this perspective, the evolution of self-conscious biological life on Earth embodies something profoundly important and sacred. It is this biological emergence that allows Awakened Awareness to recognise itself within the folds of its universal body and to develop the evolutionary structures through which an unfathomable kosmic Purpose can unfold.

Of course any human being with the basic capacity for self-reflexive consciousness can radically awaken without tracing Source's line of emanation all the way back through the kosmos because Absolute Spirit is not just the top rung of the ladder, as it were, but the very wood of which the ladder is built. As such, from a kosmically scaled radically awake perspective, and as we will see in this chapter, it is this level of biological complexity that allows humanity to embody the *first stage* in the kosmic evolutionary process where Awakened Awareness might come to realize itself on the physical plane. As we will also see as we unpack this whole teaching relating to kosmic beings, the apparent rarity of such bio-evolutionary complexity as is found on Earth *does not preclude* the existence, evolution and awakening of consciousness and energy on other planets, in other solar systems, constellations and galaxies, on subtler planes. Nor does it preclude it occurring at vastly different scales than those we are used to thinking about.

Now, when we begin to consider the ontological existence of kosmic beings (e.g. kosmically superconscious entities manifesting through universes, galaxies, constellations, stars, planets and kingdoms) from an integral perspective, we begin to get into some interesting areas. For many, things like kingdoms of nature, planets, solar systems, constellations, galaxies and universes would normally be thought of as collective structures that are composed of many parts that may or may not contain sentient, self-conscious beings. For such people, it might be hard to conceive of such structures as being literally alive, conscious entities. Theoretically, we could say that the crux of this point comes down to whether we would see something like our own planet, for instance, or a solar system or galaxy, as what Ken Wilber has called an individual or social holon. Here is how Wilber defines the difference between these two:

> The easiest way to tell the difference between an individual holon and a social or communal holon, [is that] the former has a visible physical boundary. An ant is an individual holon, an ant colony is a social holon; a human organism is an individual holon, while a family, a club, and a nation are human social holons.[155]

Wilber has been deeply critical of the confusion he sees in many models of ecology, evolution, and spirituality that do not differentiate individual and social holons. He points out that a critical consideration in whether something is classified as an individual or social holon is whether it has a single nexus of control—what he calls a dominant monad—which can direct all of its constituent parts. He illustrates this in the following way:

> The example I usually give, of why individual holons are not the same as social holons (or, why the Great Web is greatly confused), is that of my dog, Isaac, who is definitely a single organism on most days. Single organisms have what Whitehead calls *a dominant monad*, which simply means that it has an organizing or governing capacity the all of its subcomponents follow. For example, when Isaac gets up and walks across the room, *all* of his cells, molecules, and atoms get up and go with him. This isn't a democracy. Half of his cells don't go one way and the other half go another way. 100% of them get up and

---

155. Wilber, 2003.

follow the dominant monad. It doesn't matter whether we think the dominant monad is biochemistry or consciousness or a mini-soul or a material mechanism— or whether that nasty "dominant" part wouldn't be there if we were just all friends and cooperated—whatever it is, that dominant monad is there, and 100% of Isaac's cells and molecules and atoms get up and move.

And there is not a single society or group or collective anywhere in the world that does that. A social holon simply does not have a dominant monad. If you and I are talking, we form a "we", a social holon, but that "we" does not have a central "I", or dominant monad that commands you and me to do things, so that you and I will 100% obey, as Isaac's cells do. That just doesn't happen in social holons anywhere. You and I are definitely *not* related to this "we" in the same way that Isaac's cells are related to Isaac.[156]

Wilber makes a compelling point here, and yet the Trans-Himalayan teachings beg to differ. Indeed, the masters who have communicated the teachings speak very clearly about each kingdom, planet, solar system and beyond definitely being the body of manifestation of a vastly-evolved, post-human, intelligent, conscious being. According to these teachings, the structures that form such beings' bodies of manifestation (e.g. kingdoms, planets, solar systems, etc.) *do* fulfill the criteria of being individual holons and having a dominant monad, just as much as they also qualify as being social holons. In the Trans-Himalayan teachings, the planetary and kosmic Logoi who are described as incarnating through such mega-structures of manifestation as we are discussing here are in continual, vast, deep meditation. In this meditation, they hold the entire field in coherence with their moment-to-moment application of conscious Will, and transmit into it the envisioned Purpose for that field. If those beings were to withdraw their attention from the fields through which they are incarnating, the Trans-Himalayan masters suggest that the fields themselves would begin to disintegrate, just as the physical body of a human being begins to disintegrate subsequent to the withdrawal of consciousness and life at physical death. According to these teachings, such massive structures as kingdoms of nature, planets, solar systems, constellations, and beyond *do* move and shift when their incarnating entities move and shift, and therefore *do* meet the criteria to be considered as individual holons with dominant monads.

Here is an example of this in a teaching given by Djwhal Khul to Alice Bailey that speaks about the process whereby a human being,

---

156. Wilber, 2007, p. 145.

a planetary Logos or a solar Logos withdraws from their embodied sheaths at the end of their cycle of manifestation:

> The [Soul] ceases to be attracted by its form on the physical plane, and proceeding to inbreathe, withdraws its life from out of the sheath. The cycle draws to a close, the experiment has been made, the objective (a relative one from life to life and from incarnation to incarnation) has been achieved, and. there remains nothing more to desire: the [Soul], or the thinking entity, loses interest, therefore, in form, and turns his attention inward. His polarisation changes, and the physical is eventually dropped.
>
> The planetary Logos likewise in His greater cycle (the synthesis or the aggregate of the tiny cycles of the cells of His body) pursues the same course; He ceases to be attracted downward or outward, and turns His gaze within; He gathers inward the aggregate of the smaller lives within His body, the planet, and severs connection. Outer attraction ceases, and all gravitates towards the centre instead of scattering to the periphery of His body.
>
> In the [solar] system the same process is followed by the solar Logos; from His high place of abstraction, He ceases to be attracted by His body of manifestation. He withdraws His interest and the pairs of opposites, the spirit and the matter of the vehicle, dissociate. With this dissociation the solar system, that "Son of Necessity," or of desire, ceases to be, and passes out of objective existence.[157]

And again, on the solar Logos,

> Again in the solar system itself similar action will eventuate at the close of a Mahamanvantara. The Logos will withdraw within Himself, abstracting His three major principles. His body of manifestation—the Sun and the seven sacred Planets, all existing in etheric matter—will withdraw from objectivity and become obscured. From the usual physical standpoint, the light of the system will go out. This will be succeeded by a gradual inbreathing until He shall have gathered all unto Himself; the etheric will cease to exist, and the web will be no more. Full consciousness will be achieved, and in the moment of

---

157. Bailey, 1925, p. 131.

achievement existence or entified manifestation will cease. All will be reabsorbed within the Absolute; pralaya, or the cosmic heaven of rest will then ensue, and the Voice of the Silence will be heard no more. The reverberations of the WORD will die away, and the "Silence of the High Places" will reign supreme.[158]

It is of course important to acknowledge that the difference between the common human perception which says there is no way we could conventionally consider such mega-structures to be living, breathing super-conscious beings, as is embodied in Wilber's position and that put forward by the Trans-Himalayan masters. The gap between these two perspectives is an expression of the vast difference in development that exists between their relative kosmic addresses. As such, one way to consider such teachings on kosmic beings as those given by the Trans-Himalayan masters from the altitude of evolutionary awakening at which many of us find ourselves (typically personality identification, perhaps opening into soul) is to take the position that for now, these realities must remain metaphysical speculations. In order to take these ideas from metaphysical claims to what Wilber would call *post-metaphysical*, one would have to follow a particular injunction: evolve to an altitude of identification and plane access at the monadic level within the kosmic physical plane, as well as into soul and monadic forms of intelligence; gain data as a result of the experience; check that experience in an inter-subjective form of validation with other beings who have also evolved to the same level and have also collected the data. This movement corresponds to the more advanced structure-stages of the initiations previously discussed. It is at this stage that these ideas can be validated using an integral post-metaphysics. Until then they remain useful speculations framed in a trans-rational set of verifiable injunctions.

Another way to approach these teachings, which can bring the possibility of determining whether or not we consider them "trustable" a little quicker, is to ask the question of whether the beings who have communicated them are operating from a kosmic address at which such questions are determinable, even if they aren't as yet for us where we stand. If the answer to that question is "no", then we can quite happily stick with what we know already, which might suggest that this kind of thing is metaphysical New Age woo-woo. Should the answer to that question be a "yes", however, then the appropriate response on our part would be take such teachings on as eyewitness reports of reality

---

158. Bailey, 1925, p. 87.

as it exists beyond the horizon of our present sight, and to enact the responsibilities and wonder entailed in that position accordingly.

The Trans-Himalayan masters would suggest that while the current topic might involve realizations that are far in advance of what most of humanity has current access to, the path into such realizations is available to all. They would further suggest that when our decalage of vector development *does* allow the verification of these injunctions, the universe arises as *beings enfolding beings enfolding beings*, all within, through and as the matrix of Unconditioned Presence.

This is the perspective that we move from in the continuance of our consideration of the holonic kosmic ecology offered in the Trans-Himalayan teachings. In a certain sense, this basic vision of kosmic holarchy is not new. It was pioneered by the philosopher Arthur Koestler and then extended and incorporated into Integral Theory brilliantly by Wilber. Similarly, nor is the vision of holarchy as composed of infinitely extending beings enfolding beings new. That has been a regular feature of ancient esoteric cosmology, from Mahayana and Vajrayana Buddhism, to Vedanta and Hinduism more generally, to Ancient Egyptian spirituality, Ancient Greek, Pythagorean, Gnostic and Kabbalistic esotericism. Here in the Vedantic teaching of Krishnananda, we read of Ishvara, the being ensouling the universe:

> Īsvara, therefore, is everything, the coming in and the going out of all things, of all beings. Such is the glory, the magnificence and the greatness of God, Īsvara, whose integral parts, organic limbs, are the jīvas, and all things, animate or inanimate. The distinction of living and non-living beings, the inorganic and the organic, do not obtain in the realm of Īsvara's Being. For Him, it is all Consciousness. There is no jadatva, or no dead matter, for Īsvara, because it is His Being. He permeates all things....[159]

And here is an example of the Tibetan Buddhist understanding of this being, this being, described in the Tibetan Buddhist tradition as Vairochana, from Tulku Urgyen Rinpoche:

> Dharmakaya [Awakened Awareness] is all-pervasive and totally infinite, beyond any confines or limitations. This is so for the dharmakaya of all buddhas. There is no individual dharmakaya for each Buddha, as there is no individual space for each country. You cannot say there

---

159. Krishnananda, 2008, p. 79-80. Recall footnote in past chapter relating to Ishvara as a representation in both radical and evolutionary vectors.

is more than one space, can you? It is all-pervasive and wide open. It's the same with the dharmakaya level of all buddhas. That is the dharmakaya space within which the sambhogakaya manifests. No world anywhere in the universe takes form outside of the three kayas [the three state-phases of Reality]—it is simply not possible. The three kayas are the basic dimension within which all mundane worlds manifest and disappear.

This basis is also known as dharmadhatu. Out of this the sambhogakaya appears. The greater sambhogakaya is known as the Five Fold Immense Ocean Buddhas. In the middle is Buddha Immense Ocean Vairochana. To the east is immense ocean Akshobya. To the west is Immense Ocean Amitabha. To the south is immense ocean Ratnasambhava, and to the north is Immense Ocean Amogasiddhi.

The size of these greater sambhogakaya buddhas is described in as the following. The Buddha Immense Ocean Vairochana holds in his hands, in the gesture of equanimity, a begging bowl of pure lapis. Within this there is an immense ocean. Within it a tree grows and puts forth twenty-five fully opened lotus flowers. The thirteenth of these blossoms at the level of his heart, while the twenty-fifth is at the level of his forehead. We ourselves, our world, are somewhere in the thirteenth lotus at the level of the heart center. This lotus has thousands of petals and hundreds of thousands of small anthers. Within each atom of each of these anthers are one billion universes, each group being the sphere of one supreme nirmanakaya buddha.

According to the general Mahayana system, this is as far as the sphere of influence for each of the thousand buddhas of this aeon extends. Their influence is likely only one of the particles within the thousands of places within one of these lotuses. Also, in each pore in the body of Buddha Immense Ocean Vairochana, there are one billion world-systems.[160]

Both of these quotes point to how this holarchical cosmology of beings enfolding beings enfolding beings, is well-established in the wisdom traditions. What is new, and what we are offering here, is the grounding of this wisdom in a trans-rational, multi-perspectival

---

160. Tulku Urgyen Rinpoche, 1999, 103-104.

and globally oriented spirituality, enacted from an integral altitude of structural development. In taking this further, unpacking it and exploring such a worldview's relevance for where humanity currently stands today, we use Ken Wilber's Comprehensive Theory of Subtle Energies as a starting base.[161] Not every detail of where we end up will necessarily be reflective of the post-metaphysical position he outlines there, but it is one of the best considerations of the topic enacted from a post-rational structure of intelligence available, and so it is a great foundation upon which to build.

Wilber bases his theory on four orienting hypotheses:

1. *Increasing evolution brings increasing complexity of physical form.* This has been a ubiquitous finding among the sciences (physics, chemistry, biology, psychology) and displays itself in the increment of complexification demonstrated by the evolutionary process, from "quarks to protons to atoms to molecules to cells to complex organisms".[162]

2. *Increasing complexity of form (in the UR) is correlated with increasing interior consciousness (in the UL).* This stands to reason. As increasingly complex and sophisticated forms emerge (from the sub-cellular to cellular to super-cellular levels, at which we find the biological organization necessary for a limbic system and neocortex, for instance), there is the symmetrical emergence of increasingly sophisticated modes of consciousness able to operate through them.

3. *Increasing complexity of physical form is correlated with increasing subtlety of energies.* This hypothesis connects the first two. It states that with the emergence of progressively complex levels of biological organization, there emerge correspondingly subtle-energy fields surrounding them. So, whilst objects composed of atoms, molecules and ions whose condensation into the solid state of structural rigidity (such as a rock or mountain, for example) may be correlated with simply a physical body, with the emergence of the rudimentary levels of biological organization (e.g. prokaryotes, eukaryotes) there arise etheric subtle bodies surrounding these physical forms. With more complex levels of biological organization, to the point of the emergence of organisms with brainstems and limbic systems (e.g. sharks,

---

161. Wilber, 2006.

162. Wilber, 1995.

rats, and organisms of similar levels of organizational complexity), there arise astral-emotional subtle bodies, transcending yet including the etheric and physical. With the continuation of that trend of complexification, beings with neocortices and the capacity for increasingly sophisticated cognitive and affective operations emerge (such as humans and dolphins) whose forms are correlatively surrounded by mental subtle-bodies (thought-fields) that transcend and include the astral, etheric, and physical bodies.

4. *Complexity of physical form is necessary for the expression or manifestation of both higher consciousness and subtler energy.* This is a crucial component of the overall theory. This hypothesis qualifies previous hypotheses by relativizing the claim that such subtle bodies *emerge* with the complexification of matter. Rather, this point incorporates the teachings of the traditions on the involutionary as well as evolutionary processes Spirit passes through on its passage into and out of the physical. It thus recognizes that there very well may be beings operating on the etheric, astral or mental planes right now (as well as on every other plane and on every octave of the kosmos) who lack any physical form constituted of the corresponding degree of biological complexity, but that in order for those beings to *manifest* in the physical, they require such forms.

If we take the notion of kosmic beings manifesting through all octaves of the kosmos on as a hypothesis, we might suggest that for each octave of holonic entity (human, planetary, solar, constellational, super-constellational, galactic, universal) we need a new scale of content in the Four Quadrant matrix. Such content would be reflective of that being's own respective interiorities of consciousness; exterior systems of form complexification; the collective culture their consciousness forms part of; and exterior systems of organization employed by the collective 'community' of kosmic beings. If we were to explore what such a matrix would look like in the UL and UR for the planetary Logos of the Earth, for instance, we would now see that in the UR, entire kingdoms of nature (geochemical, biological, fauna, human...) serve as the exterior stages of form complexification within its physical body.[163] These support increasingly inclusive interiorities of its

---

163. Recall that just as for a being incarnating at the intra-planetary octave as a human being, their physical body is found composed of the matter of the three densest sub-planes of the physical plane, so also for a being incarnating at the planetary

consciousness and subtle energy bodies (physical, etheric, astral and mental, respectively) just as the emergence of a reptilian brain stem, a limbic system and a neocortex has done for human beings.[164]

Additionally, if we incorporate the Integral understanding that one definition of the chakras relates to stages of development, the eventual destiny of humanity, as the fourth major life-sphere of evolutionary complexity, and when fused with Hierarchy, is to express the planetary heart chakra (the fourth chakra of the seven whether we are counting from up or down). Furthermore, the above hypotheses allow us to recognize that a) there are likely deeper levels of consciousness/energy than currently found on Earth waiting for the sufficient complexification of exterior form before they can manifest (the communities of the soul and monadic planes—Hierarchy and Shamballa—which are ourselves as well as other soul and monadic awakened beings not yet manifesting in the physical), and b) that as human beings together pioneer into the soul-level uncharted territories of evolutionary novelty that are emerging on the basis of the mental-level foundation, the *planetary soul* will be increasingly enacted through complexifications of form as yet unknown to us.[165]

If we expand this vision further out into the kosmos, then our perspective holds space not just for the planetary Logos of the Earth, but for the existence of other planetary Logoi ensouling other planets, and whatever ascending levels of evolutionary complexity reside in subtle energy-matter or dense form within their planetary fields; solar Logoi manifesting through solar systems, the planets they contain, and whatever evolutionary life-spheres may reside on their sun(s) and planet(s); constellational Logoi manifesting through constellations,

---

octave is their physical body composed of matter of the three densest sub-planes of the kosmic physical plane.

164. Readers wondering if the integrity of the individual and collective quadrants has been violated by the supposition that 'kingdom collectives' form the exterior forms of a super-individual should refer to the previous footnote, and remember also the holonic organization of the regularly cited exteriors. Organisms are composed of cells, which are composed of molecules, which are composed of atoms, which are composed of...

165. Contemporary evolutionary neuroscience suggests that the prefrontal cortex is the most recent phylogenetically developed brain structure to develop, which, in company with the rest of the cortex, the limbic system and brain stem, supports our psychic, cognitive-affective consciousness. What neural structure might come next to support the subtle planetary soul is a fascinating question. We suggest it may not be a new region of neural tissue *per se*, but rather the much fuller integration of the already extant neural regions. Recent research on the neural signatures demonstrates by advanced meditators supports this hypothesis.

which may contain solar systems, planets, and evolutionary life-spheres perhaps operative in subtle or dense energy-matter on all three; super constellational Logoi manifesting through systems of constellations, which may contain solar systems, planets, and evolutionary life-spheres on those kosmic bodies; galactic Logoi incarnating through galaxies, perhaps containing constellational systems, solar systems, and planets; all within the super-holon of the universal Logos manifesting through the entire universe of galaxies, super-constellations, constellations, solar systems, planets, and whatever evolutionary life-spheres may exist at all of those levels in subtle energy-matter or dense form.

According to this approach, every level of holarchy is as much a living, conscious entity, an I, as it is capable of being understood as IT, part of a shared cultural WE, and a system of dynamic energy (ITs).[166] Furthermore, we could say that each level of holarchy functions on two basic energy streams: eros and agape. Eros drives the being to higher levels of self-organization, complexity, and more embracing life-spheres, while agape motivates their reaching downward to care for their constituent parts that compose and support it—to hold a field for their evolution and awakening out of loving sacrifice.

Thus, when the holonic kosmos is enacted from the root vantage point of Awakened Awareness, a deep enough level of identification, a wide enough spectrum of recognized relationship, a full enough attainment of plane access and soul to monadic forms of intelligence, all individual holons possess all Four Quadrants, all the way up, and all the way down—and this isn't just a conceptualized understanding, it is a living *experience*. The Earth itself can therefore be understood as the exterior body of manifestation (the IT in the UR Quadrant) of a kosmically awakened Life (the 'I' in the UL Quadrant) that is part of a kosmic community of such Lives (the WE of the LL Quadrant), and whose body is constituted of lives that have the opportunity to participate in its Life (the ITS of the LR Quadrant). Similarly, from this decalage of awakening, identification, relationship cultivation, plane access and intelligence unfoldment all human monadic essences, other planets, solar systems, constellations, constellational systems, galaxies and potential universes can be considered in the same way. That is, each can be considered as an 'I' that is manifesting through 'IT', that is joined in a 'WE'. And at each octave, the kosmic being reaches ever forward (eros) into further horizons of kosmic infinity,

---

166. Furthermore, from an integral perspective we must note that groups/systems of individual holons, enacted only as social holons from a personality, soul or monadic level of identification within the kosmic physical plane (e.g. galaxies), seem to flip to individual holons when enacted from a level of identification on more subtle kosmic planes.

whilst simultaneously holding a field for the evolution and awakening of all constituent beings that compose it (agape).

Furthermore, the above four hypotheses of Wilber's, when taken together, allow us to begin to see what an extraordinary space humanity occupies within the overall planetary life and kosmos. Humanity is the first stage of the evolutionary process at which self-reflective consciousness has complex enough forms (very likely the level of thalamo-cortical reciprocity in information processing we demonstrate) to support it. It is therefore the first stage at which access to all three altitudes of identification on the kosmic physical plane—personality, soul and monad—is possible. Incredibly, as discussed above, and as has been the testimony of the siddhas and realizers of the ages, it is thus the first point on the physical plane at which the Absolute may come to recognize its own Original Face.

Naturally, as our recognition of this unfolds as a collective planetary culture and as our scientific explorations into the kosmos proceed apace, we wonder "Are we alone in this? Are there other civilizations in the kosmos where Absolute Reality is awakening to itself, or has done in the past or will do in the future?"

Now, at this point, at this exciting time in the history of astronomical and evolutionary biological research during which our knowledge of the universe is expanding so rapidly, we are forced to entertain two fascinating hypotheses:

H1: *Despite the astonishing frequency with which exoplanets and super-Earths are being discovered every day, the existence of biological life throughout the universe is rare.*

H2: *As the astonishing frequency with which exoplanets and super-Earths are being discovered every day suggests, the existence of biological life in the universe may not be rare.*

The first of these hypotheses is in alignment with the view held within the scientific community until about a decade ago concerning the existence of extraterrestrial biological life. The second is in alignment with the view increasingly prevalent within the astronomical community today, as expressed by the esteemed astronomer and former president of the Royal Astronomical Society, Sir Martin Rees.[167] If the first is true, within the context of our current considerations, we would be forced to suggest that while every octave of the universal field has its corresponding consciousness (the planetary, solar, constellational, galactic Logoi), the orientation of those consciousnesses *is not physical*

167. Blake, 2010.

plane focused. This conclusion would stem from the atomic, molecular, liquid and gaseous structures of the majority of celestial bodies (planets, suns, galaxies…) that could likely not support biological life as we understand it.

In exploring this in the light of the above four hypotheses, we recall that just because a phenomenon (such as kosmic civilizations or beings—planetary, solar, stellar…) may not disclose itself to physical plane observation, nor embody itself in the evolution of life on the physical plane, *this does not mean that it does not exist on subtler levels.* Rather, in such a case where we have other pointers to the possibility that they do exist (such as information returned from awakened penetration into the truths and Truth residing on the soul, monadic and even subtler kosmic planes) we instead take the position that the complexification of forms necessary for their incarnation on the physical plane is not yet present. In the light of the multidimensional vision of the kosmos that we have been exploring in this book, this should not surprise us. There are many planes upon which life can be evolving, and heaven forbid we should get physical plane-centric!

Under this hypothesis, biologically self-consciousness humanity is something extraordinarily rare and precious. In the words of Djwhal Khul, "Human civilisation is a rare commodity in the kosmos. A human civilisation awakened as Love is rarer, and one awakened to Life is a jewel."[168] In the Trans-Himalayan teachings, the planetary Logos of the Earth is spoken of as a "Divine Rebel" that has chosen to attempt to anchor the awakening of the One Absolute Life—Awakened Awareness—to itself *through humanity* at a level of physical plane density in material incarnation only rarely before attempted in the kosmos. This relates to the suggestion that the reason the vast majority of other planets do not disclose themselves readily as spheres evolving through the development of biological life is that the creative evolutionary focus of their ensouling entity is not presently physical plane oriented. The wonder of this possibility is captured beautifully in Djwhal Khul's work with Alice Bailey:

> The confines of the Heavens Themselves are illimitable and utterly unknown. Naught but the wildest speculation is possible to the tiny finite minds of men and it profits us not to consider the question. Go out on some clear starlit night and seek to realize that in the many thousands of suns and constellations visible to the unaided eye of man, and in the tens of millions which the modern telescope reveals there is seen the physical manifestation of as

168. Lyon, 2010, p. 74.

many millions of intelligent existences; this infers that what is visible is simply those existences who are in [gross realm] incarnation. But only one-seventh of the possible appearances are incarnating. Six-sevenths are out of incarnation, waiting their turn to manifest, and holding back from incarnation until, in the turning of the great wheel, suitable and better conditions may eventuate.

Realise further that the bodies of all these sentient intelligent cosmic, solar and planetary Logoi are constituted of living sentient beings, and the brain reels, and the mind draws back in dismay before such a staggering concept. Yet so it is, and so all moves forward to some unfathomable and magnificent consummation which will only in part begin to be visioned by us when our consciousness has expanded beyond the cosmic physical plane, and beyond the cosmic astral until it can "conceive and think" upon the cosmic mental plane. That supposes a realisation beyond that of the Buddhas who have the consciousness of the cosmic physical plane, and beyond that of the planetary Logoi. It is the consciousness and knowledge of a solar Logos.

To the occult student, who has developed the power of the inner vision, the vault of Heaven can therefore be seen as a blazing fire of light, and the stars as focal points of flame from which radiate streams of dynamic energy. Darkness is light to the illumined Seer, and the secret of the Heavens can be read and expressed in terms of force currents, energy centres, and dynamic fiery systemic peripheries.[169]

If the second hypothesis is true—that biological life may *not* be nearly so rare as we formally conceived—then our view of the universe and our understanding of ourselves within it takes what some theorists have suggested to be the greatest shift in known history. Going by the current science, if we factor in the astrobiological supposition that any planet on which biological life was able to evolve would need to be within the habitable, or "goldilocks zone" at which distance from its central star liquid water would be maintained on its surface, then there would therefore likely be one planet per exosystem with the capacity for biological life. Incredibly, if we do the math, in a galaxy such as our Milky Way in which there are an estimated 200-400 billion stars, that makes for a massive number of potential other planets on which biological life

---

169. Bailey, 1925, p. 1059-1060.

could evolve. Furthermore, owing to ongoing findings within the field of evolutionary biology, it is now increasingly understood that simple biological life can exist in some of the harshest environments known. The criteria by which scientists deem certain environments likely to be inhabitable by biological life are shifting rapidly, and some scientists have begun to speculate that such life may be far more prevalent throughout the kosmos than previously considered.[170] With more and more scientists and researchers seriously considering that there is no conceivable reason why that life, if it exists, should be less evolutionarily advanced than it is on Earth, but may in fact be vastly more advanced in some cases, our entire understanding of our humanity could be on the verge of a radical shift.

In the Trans-Himalayan teachings, the term 'humanity' is not tied exclusively to what we understand by that term on the Earth, but is rather used for all self-conscious manifestations of a planetary Logos wherever found in the kosmos. According to the masters of this teaching, self-consciousness does not exist in the pre- or post-human evolutionary life-spheres, wherever they arise in the kosmos, in the same way. In this sense, the orientation previously discussed toward anchoring the realization of Awakened Awareness to itself at a deeper point of physical plane immersion than attempted before remains, but instead of it being an adventure engaged only by Earth humanity, it is one engaged by *kosmic humanity*.

---

170. Arsenic Loving Bacteria may Help in Hunt for Alien Life, 2010.

Chapter 19
Transmitting Purpose
Jon

Having considered the initiations, kosmic paths and kosmic Logoi as we have, it is worth us exploring the point that the path is not merely unidirectional. The kosmic path we articulate does not ascend only from smaller to greater octaves of the kosmos, or from planetary to solar to constellational to super-constellational to galactic to universal spheres of identification and incarnation. The smaller octave expressions of evolution in the kosmos (such as a planet like Earth, for instance, and all the lives upon it) are those in which the larger kosmic beings, in whose bodies that sphere exists as a constituent holon, are able to begin to fully incarnate their vast levels of awakening and development. It serves us to recognize, therefore, that an increase in kosmic scale/octave does not *necessarily* entail more developed expressions of awakening, identification, relationship, vision and, intelligence.

In fact, according to the Trans-Himalayan transmission, some of the greatest movements of love and enlightened action in the kosmos occur when Great Beings choose to incarnate in the most dense, compacted and materially imprisoned octaves of manifestation. We might imagine such beings as those who have retraced their line of radical freedom, identification and evolutionary development into vast scales of kosmic liberation, perhaps at the stellar or even galactic octaves, but who then wilfully sacrifice those heights so that radical awakening, divine identification, relationship, vision and intelligence can be unfolded at a new level of kosmic evolutionary complexity. It is with this in mind that we bring our attention to Earth and consider its relevance and unique place in the larger kosmic ecosystem. To explore this, we venture further into the freedom of incarnational expression that is the property of a master, the kosmic planes and octaves of body through which such a being might incarnate, and into the subject matter of this chapter—the transmission of Purpose.

Subsequent to the fifth initiation and entry into the post-mastery initiations of the kosmic paths, the enlightened being stands so free of exclusive identification with any aspect of manifest reality that, for the first time, there enters the factor of *choice* into their decision concerning what comes next. This choice is determined by their pre-established depth of awakening, identification, relationship cultivation, plane access and intelligence, in the kosmos. Crucially, what has most relevance in this chapter is the opportunity that faces such a being to hold *different, asymmetrical levels of identification and plane abiding in their chosen sphere of incarnation.* One side of this asymmetry arises through their identification with and as the identity, envisioned Purpose, and Life-force of the greater kosmic Logoi within whose bodies they set up their own holonic field of manifestation, taking responsibility for the evolution and awakening of all life therein. The other side arises through their wilful, shamanic penetration into their chosen plane of expression and service, which may be on a much denser plane than the where the height of their relative identification is rooted. From their chosen plane of incarnation, they receptively open to the kosmic Wakefulness, Identity, Purpose-transmission and energies of the kosmic Logoi of whom they are emissaries at more densely incarnated levels of the universe. In order to unpack this teaching more clearly, as was noted in the previous chapter, we track seven different octaves at which kosmic Logoi are incarnating, with each of these having personality, soul and monadic sub-levels of bodymind expression. To illustrate this we reproduce here figure 27:

Octave	Incarnating Entity
1. Universal	Universal monad
2. Galactic	Galactic monad
3. Super-constellational	Super-constellational monad
4 Constellational	Constellational monad
5. Solar system	Solar monad
6. Planetary	Planetary monad
7. Human	Human monad

Table 5. Table showing the seven octaves, or scales of incarnation, through which kosmic Logoi, each with various degrees of evolutionary development, are taking form.

In relation to this table it should be noted that each structure-stage transition from body to personality to soul to monad at each octave resonates, reverberates, and holds the capacity for identification with the correlating structure-stage at other octaves. This means that a human buddha, radically awake and monadically identified, is able to cultivate identification with the monadic bodymind, envisioned Purpose, and electric Life-force of a planetary, solar, constellational, super-constellational, galactic or eventually the universal Logos. This is the great life known in the Indian Hindu and Tibetan Buddhist cosmologies as Ishvara and Vairochana respectively. Additionally, it is important to note that although we point to seven octaves here, we are open to the possibility that the altitudes of identification continue indefinitely. From the universal octave they may move to the multi-universal and beyond; each octave containing a bodily, personality, soul, and monadic dimension.

Now, as it will hopefully have become clear over the course of this book, when we speak of Purpose we are speaking of it not in terms of an envisioned goal or concept that is something other than the one holding it. The transmission of Purpose is an innate quality of the monad at all octaves of manifestation, with that Purpose being the unique and most powerful expression of being and service, both rooted in and in service to the Totality, which it is the honor of each being to live. For a human being, this only begins to come online once they begin to contact the monadic depth of identification, but for liberated masters and buddhas this is the *root* from which their radical awakening to Boundless, Selfless, Conscious Light and their evolutionary awakening—or the kosmic re-tracing of the root of their identification—expresses in kosmic service. In the words of Djwhal Khul and Bruce Lyon, and in relation to our capacity for identification with the planetary Logos of the Earth:

> The deep truth at the core of our being is that 'we hold the universe in our embrace and the galaxies are our children'. If this is so we might at least be able to sustainably hold identification with the level of shared responsibility for one small planet. When we can do this, what is revealed to us is the One who has already taken that responsibility and therefore is in a position to share it with us.[171]

From a bottom-up perspective, when beings are developing and unfolding within planetary fields such as the Earth, those fields serve as schools or training grounds which may prepare such beings for mature life in the open field of the kosmos. When the kosmic field is

---

171. Lyon, 2007, p. 44-45.

entered, just as is the case when an adolescent leaves high school or university, the safety bars are taken away and the stakes are higher. Now there are greater responsibilities, greater risks, and greater rewards. When buddhas work from the monadic altitude, they work through identification with and as the greater, more inclusive identity of a higher kosmic Logos. There, such buddhas have assumed responsibility for a unique expression of the kosmic Purpose envisioned by those Logoi incarnating at greater octaves. As such, their bodhisattvic and avataric kosmic service can be understood as an unbroken identification with and as a kosmic Logos incarnating at a greater octave. This allows the lightning-transmission of the latter's envisioned kosmic Purpose and monadic energy, or Life-force, into the former's field of incarnation, where it works *dynamically*.

It is this dynamic transmission of kosmic Purpose and Life-force into a Logos' field of incarnation that *automatically* calls forth the evolutionary revelation of the Absolute within and through its entire radius of life, as well as the unfoldment of its unique expression of kosmic Purpose. Crucial here is the identification between a Logos at one octave of incarnation (either human, planetary, stellar or galactic, for instance) with a particular altitude of identification (personality, soul, monad) of a Logos at a greater octave of incarnational expression. This allows the latter to act as a dynamo for the radical and evolutionary awakening of all in the former's field of awakening, *according to a scale of significance determined by the latter.* An example could be the identification of a planetary Logos abiding on the kosmic astral plane, with the identity of another, more all-encompassing being, such as a galactic Logos abiding on the kosmic monadic plane. Through such identification, the former *automatically* calls forth, though on a time scale of billions of years, the radical awakening and evolutionary 4-line transmutation, transformation and eventual transfiguration of everything within its body and sphere of incarnation to a *galactic* octave of Purpose.

This is something of the *power* expression of a Logos, awake as Absolute Light, to transmute, transform and transfigure the kosmic plane and body of manifestation it is incarnating through. It does this by meditatively holding its attuned identification at one with a wider, more all-embracing life-sphere. An evocative symbol for the effect of such a Logos on its environment is the black hole. Just as a supermassive black hole is able to hold the field for an entire galaxy of one hundred billion suns to be birthed, grow, shine radiantly and eventually die, simply through the dynamic power of its infinite density at the singularity (according to Einsteinian physics), the identification of a kosmic Logos is supremely dynamic. From their pre-established base as Absolute Awakened Awareness and their monadic identification, they effortlessly evoke the awakening, evolution and unfoldment of all beings within their field. Just as a black hole draws all toward itself with titanic power, so also does the presence of Purpose within any particular field result in the gradual warping of space and time so as to conform to that Purpose.

In exploring this analogy, Djwhal Khul has taught that:

> In the same way that a gravity field is a result of the warping or influencing of space-time by the presence of a dense mass, so a field of consciousness is a result of the 'warping' of buddhic substance by the presence of a 'dense' purpose. A black hole has no form or light but large 'mass'. Purpose is related to mass, not matter. A Master in the center of an ashram is the embodiment or rather 'antiembodiment' of purpose and therefore creates a field of consciousness which is really the disruption of a buddhic field that then warps around this central mass.[172]

So, the mass of a black hole corresponds to the scale of Purpose it is holding a field for. While every black hole, according to Einsteinian physics, harbors a singularity of infinite density at its core (analogous, we might say, to the Nondual vantage point of Awakened Awareness that resides at the core of the monad), *their mass can vary widely*, all the way from being of about 10 solar masses through to the supermassive kind that can be of billions of solar masses. With this increase in mass comes an increase in the strength of their gravitational pull, and thus a corresponding increase in the number of objects (whether floating gasses, space debris or planets and suns) orbiting them. It is this mass that has the power to warp space and time so as to hold a field for billions of suns and planets, and it seems likely that the identification and Life-force of a planetary, solar or galactic Logos operates in an analogous manner.

Now, whilst the octave of incarnation such a being chooses to ensoul does not *necessarily* increase with their advancing development, the scale of kosmic Purpose and responsibility with which they are able to identify (capacity) *does*. What this means is that it is possible for a kosmic bodhisattva operating from their radically Awakened Heart to identify with and take responsibility for a unique expression and unfoldment of Purpose in the kosmos *at any octave that their octave of identification already contains*. So, a kosmic Logos whose awakening, identification, cultivated relationship, plane access and intelligences had unfolded to the stellar octave is able then to choose to ensoul any scale of the kosmos within or as a solar system (perhaps incarnating as a human avatar, a fiery meteor, or a sun, for instance) for the evolution

---

172. Unpublished Mercury transmission.

and awakening of life therein (though with the possibility of that Logos' expression of its consciousness on the physical plane being dependent on the presence of the needed level of bio-evolutionary complexity, as previously stated). Similarly. we could say that a kosmic Logos whose radical awakening, identification, cultivated relationship, plane access and developed intelligences had unfolded to the galactic scale would be able to choose to ensoul any scale of form in the galaxy.

This shines light on Bruce Lyon's speculation that the planetary Logos of the Earth may actually be a being whose kosmic radical and evolutionary awakening already extended to galactic octave, but who, from the power of identification with galactic Purpose and pure kosmic bodhisattva activity, chose to ensoul a planet at the darkest, most materially embodied and imprisoned octave of the kosmos. That being thus holds a point for the profoundly significant occurrence of the evolution and awakening of life to Absolute Reality, and the unique expression of kosmic Purpose that it is the role of the Earth to give to the galaxy. This is something we will explore in Part 7 of this book.

At all octaves of the kosmos and on all kosmic planes, beings can be more or less aligned with progressively wider scales of Purpose. However, none of the information we can glean from observation or intuition on the level of their evolution or awakening as far as matter or consciousness is concerned can necessarily tell us about what *scale* of kosmic Purpose and Life they are more or less identified with. Extending a point made in the last chapter, from one perspective we could see planets as the root and galaxies as the crown of the kosmos. This is because the scale of awakening, identification, relationship, plane access and intelligence of a galactic Logos might appear to far transcend that of a planetary Logos, and certainly any being on that planet. From the opposite perspective we could see galaxies or the universe itself as the root of the kosmos, and physical plane solid planets as the crown. This is because it is only on the latter (as we presently understand it) that the required levels of bio-evolutionary complexity necessary for the revelation of Absolute Reality and the full incarnation and expression of the solar, galactic and universal monad(s) on the physical plane.

Both of these perspectives miss something, however, and this relates to the awakened freedom inhabited by the kind of profoundly awake and evolved Logoi we are speaking of, and of the kosmic field itself. It is quite possible for a planetary Logos, such as that which ensouls the Earth, to be holding a galactic octave of identification and Life-force but to have *chosen* to express that identification through a planet on the densest octave of the kosmos. All of this relates to the point that a greater octave of incarnation does not necessarily predict a greater degree to development. A kosmic Logos can incarnate through any scale of incarnation smaller than the octave of kosmic Purpose they have developed identification with. Indeed, and as we shall later explore, this relates to the mystery communicated in the Trans-Himalayan transmission of why one appellation of the planetary Logos of the Earth, as it is known to the liberated bodhisattvas and masters of Hierarchy and Shamballa, is "the Great Sacrifice".

Earth is Eden

Chapter 20
Kosmic Community
Dustin and Jon

In the previous chapters, the context within which our planetary narrative is unfolding has been expanded into the kosmic. Considerations of vastly evolved and radically awake beings, holding planetary, solar, constellational, super-constellational, galactic and universal fields for the evolution and awakening of all life upon/within them, calls us to reassess thoroughly the way in which we understand ourselves, our planet, and the kosmos itself. This is no less the case as we consider the possibility of other humanity's and forms of life throughout the universe also.

As the pace, scale and intensity of our current discussions expands, it feels important to drop our anchor into the recognition that far from being new age flights of fancy, on deeply pragmatic levels these matters have *substance*. Considerations of kosmic beings incarnating through various octaves of incarnational fields are not merely speculative. They are drawn from and informed by the profoundly detailed explorations of these topics by master-teachers whose extraordinary levels of wakefulness, stabilized identification, multidimensional relationship, plane access, and wisdom-intelligence has allowed them to penetrate into spheres of realization in which these matters disclose themselves. In relation to our discussion of other humanities and forms of life elsewhere in the universe, our approach is rigorously informed by the theories and findings that are emerging with startling pace from the emergent scientific field of astrobiology.

Astrobiology, formally known as exobiology, is a field that has emerged from what must be one of the fastest and most significant paradigm shifts in scientific history. Only a decade ago, and as is reflected in the lag between the present perspectives within the scientific community at any particular time and those of lay people in general society, this topic was deeply controversial. Open musings on the possibility of extraterrestrial biological life were still in great part anchored in cultural taboo, and most researchers steered well clear of them. Now, though, things could not be more different. Not only is extraterrestrial biological life now one of the central points of discussion for the best scientists in the world, but the race is on as to *who will detect it first.*

This degree of surety is built upon the foundation of some of the fascinating scientific findings that have been emerging over the last decade or so. According to recent analysis of data gathered from the Kepler Space mission, for instance, each of the approximately 100 billion stars in our galaxy has at least 1.6 attendant planets circling it, making the probable number of "exo-worlds" greater than 160 million. According to Arnaud Cassan of the Paris Institute of Astrophysics, the primary author of the study referred to above:

> This statistical study tells us that planets around stars are the rule, rather than the exception. From now on, we should see our galaxy populated not only with billions of bright stars, but imagine them surrounded by as many hidden extrasolar worlds.[173]

Other startling biogenetic research has found that the formation of the first ten amino acids involved in the synthesis of proteins from genetic information stored in DNA, and decoded by mRNA so as to eventually form living cells, is not only probable at low temperatures and pressures, but that its statistical probability corresponds to the concentration of these fundamental building blocks of life as discovered on meteorites.[174] This is an extraordinary finding. It means that the ingredients necessary for simple biological life may not be a miracle at all, but that, given the degree to which exoplanets may be struck with reasonably sized asteroids with these basic ingredients upon them, the universe may be teaming with life.

It is for these reasons that the astrobiologist Dr. David Darling wrote in his book, *Life Everywhere*, that the old paradigm, which views biological life as we know it on Earth as some random miracle, may be on its way out. Darling suggests that on the basis of the current evidence, in place of this a new paradigm may emerge in which it is

---

173. The Milky Ways' Alien Planets: 160 Million and Counting!, 2012.

174. Ridley Scott's "Prometheus" Suggests DNA May Be a Constant in the Universe --Richard Dawkins and Other Scientists Agree, 2012.

understood that as long as the basic, common ingredients are present, biological life may actually be *inevitable*. Dr. David Darling:

> An increasingly common claim among researchers is that life may arise inevitably whenever a suitable energy source, a concentrated supply of organic (carbon-based) material, and water occur together.
>
> These ingredients are starting to look ubiquitous in space. Comets, in particular, are increasingly seen as significant vehicles for delivering water and organic cocktails to infant worlds. And with the discovery on Earth of meteorites from Mars, the *interplanetary* transfer of biochemicals or even life itself has become a respectable topic of debate.
>
> Both within and beyond the solar system, the list of potential places where life may have become established is growing fast. Close to home, Jupiter's moons Europa and Ganymede have taken on biological interest with the realization that they may harbor chemical-rich oceans of water beneath their icy surfaces. Farther afield, the finding of dozens of extrasolar planets, almost as soon as we knew how to look for them, encourages scientists to think that planetary systems around stars are the rule rather than the exception. From origin of life studies to complexity theory, from extrasolar planet detection to work on extremophiles, from pre-Cambrian paleontology to interstellar chemistry, the emerging message is clear and virtually unanimous: extraterrestrial life is there for the finding.[175]

This emerging science is calling us to dramatically reconsider our place in the universe, and to do it *fast*. Indeed, Jocelyn Bell Burnell, an astrophysicist from Oxford University who discovered the first pulsars in her thesis work with Anthony Hewish, who went on to share the Nobel Prize in Physics, recently stated: "I do suspect we are going to get signs of life elsewhere, maybe even intelligent life, within the next century."[176]

Burnell's point that any extraterrestrial life we discover may be intelligent is rooted in the data that is emerging on just how long

---

175. Darling, 2001, p. xii.

176. "Evidence of Alien Life Expected Within 21st Century" --Leading Astrophysicist, 2012.

biological life might have existed on exoplanets, should it have begun to evolve in the first place. Until recently it was believed by the scientific community that rocky, terrestrial planets such as Earth, which are made up largely of silicate rocks or metals, would not have been able to form until comparatively recently in the history of the universe. This was because the fundamental ingredients of the rocks from which they are composed—chemical elements such as silicon and oxygen—required such large time scales of nuclear fusion in the hearts of suns before being produced. Recent research, however, has detected the presence of planets orbiting suns whose composition deeply challenges this, leading scientists to suggest that rocky planets upon which biological life could evolve may have existed for far longer than we previously conceived.[177] This necessarily means that the biological life itself may have had ample time to evolve into forms we would classify as 'intelligent'.

It is important to make the point that this is not 'fringe' science. It is mainstream science; so mainstream, in fact, that some of the biggest and most reputable research organizations are, so to say, putting their money where their proverbial mouths are. Here again, Dr. David Darling:

> The dominant question is where the search for extraterrestrial life should be focused here and now. And the answer is obvious from every astrobiological program, underway or planned, in which there is a significant investment of funds and other resources. It's evident in the 'Roadmap' drawn up by the astrobiological Institute to help guide NASA's activities in this field. It's evident in the overwhelming majority of papers published on the subject of extraterrestrial life in leading scientific journals and in the proceedings of relevant conferences, such as the first annual science conference on astrobiology, held at the Ames Research Center in April 2000. Most tellingly, it's evident in the design and implementation of the multimillion dollar instruments that have been built, or are being built, to test for the presence of biological activity on other worlds.[178]

So, with the astrobiological field is pioneering out into the kosmos with its penetrative eye, and many of the most respected scientists in the world declaring it is only a matter of time (and likely quite a short

---

177. Alien Habitable Planets May Exist Billions of Years Older Than Earth --Harvard Center for Astrophysics, 2012.

178. Darling, 2001, p. 12-13.

time) before extraterrestrial biological life is detected, these matters are deeply salient. But what if we have already experienced contact with life from other worlds? What if, before we ever got so close to being able to detect them, *they were already here?*

Contact

As we step into an exploration of the possibility contact with extraterrestrial beings from other worlds—other humanities as we have touched on in previous chapters—it is important to note that inquiries like these pose massive challenges to the prevailing worldview of our time, and the way we conceive of reality itself. We noted earlier that it is possible that exoplanets beyond our solar system may actually have existed far longer than either the Earth, or what we previously considered possible. Indeed, Milan Cirkovic at the Astronomical Observatory in Belgrade has pointed out that the median age of rocky, terrestrial planets in the Milky Way is approximately 1.8 gigayears (1 gigayear = 1 billion years) older than our solar system and Earth. That means that the median age of any extraterrestrial civilizations that had arisen on those planets and continued to survive could be up to 1.8 gigayears older than human civilization.[179] To put this vast amount of time in perspective, the entirety of what science understands as human evolution amounts to about 1 million years. It therefore seems quite reasonable to consider seriously a) the existence of extraterrestrial civilizations that may be *far in advance of our own*, and b) that according to a Four Quadrant understanding of the way that consciousness, biology/behavior, culture and civilization/technology tetra-arise and evolve together, any such civilizations that had developed the technological capacity to travel to Earth would likely be far more evolved than Earth humanity. This supposition is in accordance with the Trans-Himalayan transmissions from the awakened beings of Hierarchy. As Djwhal Khul tells us:

> There are many galaxies in cosmos that do not even have one planet, in which the forms and the consciousness have reached the level of evolution that they have on Earth. Other galaxies have many planets that have evolved far beyond Earth.[180]

---

179. World-Leading Physicist Says Extraterrestrials "Could Exist in Forms We Can't Conceive", 2010.

180. Lyon, 2010, p. 76.

Arthur C. Clarke once noted that any sufficiently advanced technology would necessarily be indistinguishable from magic. Similarly, the world-renowned physicist Sir Martin Rees, of Cambridge University, and astrobiologist Paul Davis, of Arizona State University, recently posed the question of if we were to encounter alien technology far in advance of our own, would we even understand what it was? For Sir Martin Rees, speaking at a recent conference, this goes also for the alien beings themselves:

> They could be staring us in the face and we just don't recognize them. The problem is that we're looking for something very much like us, assuming that they at least have something like the same mathematics and technology.
>
> I suspect there could be life and intelligence out there in forms we can't conceive. Just as a chimpanzee can't understand quantum theory, it could be there as aspects of reality that are beyond the capacity of our brains.
>
> I certainly think that humans are not the limit of evolutionary complexity. There may indeed be post-human entities, either organic or silicon-based, which can in some respects surpass what a human can do. I think it would be rather surprising if our mental capacities were matched to understanding all the levels of reality. The chimpanzees certainly aren't, so why should ours be either? So there may be levels that will have to await some post-human emergence.[181]

Any civilization that was able to travel here would first have to navigate successfully the extreme technological challenges of such vast distances of travel. In the following, we suggest two hypothetical technologies that might account for the possibility of extraterrestrial entities being able to travel to Earth. The first is a technology of matter. The second is a technology of consciousness.

A Taxonomy for Classifying Kosmic Civilizations

In 1964, the Russian astrophysicist Nikolai Kardashev suggested a taxonomy for classifying extraterrestrial civilizations according to how they source their energy. Kardashev's scale is still drawn upon today. For Kardashev, Type 1 civilizations would harness their energy from their

---

181. Are Humans the Limit of Evolutionary Complexity?, 2009.

own planet, like we do on Earth, primarily through the extraction of natural resources such as coal, gas, and oil. Type 2 civilizations would harness their energy from the star their planet orbits, allowing them to increase the amount of energy they would be able to draw upon by a factor of 10 billion. This is something we see the beginning of on Earth in relation to solar power. Type 3 civilizations would harness their energy from their entire galaxy, again raising the amount of energy they were able to drawn upon by a further factor of 10 billion.[182] For reasons that remain subject to investigation and further research, black holes periodically emit bursts of intensely powerful electromagnetic radiation called gamma rays. Indeed, these gamma rays are so powerful that they can release more energy in 10 seconds than our sun will release during the entirety of its 10 billion year lifetime.[183]

In a recent paper titled *"Black Holes: Attractors for Intelligence",*[184] Clement Vidal, from the Evolution, Complexity and Cognition group at the Vrije Universiteit Brussel, even went so far as to suggest that a stage beyond this would involve such a civilization's capacity to manipulate the physical reality of space-time itself. Human scientists today have learned to manipulate atoms using nanotechnology, and Vidal suggests that a civilization far in advance of our own might have developed their technological expertise so far as to create their own black holes from which to harness energy at any time, anywhere. In connection to the possibility of extraterrestrial civilizations developing material means enabling kosmic travel, it seems reasonable to hypothesize that for any civilizations that had moved into the Type 2 or Type 3 category, navigating the technological hurdles encountered in traveling the distance to Earth would become a definite possibility.

In relation to a consciousness-based technology for travel, we feel that some of the ageless wisdom held in the Earth's esoteric traditions may have something to offer here on the kind of subtle travel and voluntary abstraction or manifestation into the physical. We particularly look to the highly esoteric and advanced phenomenon of what is known in the Tibetan tradition as Rainbow Body, or in the Trans-Himalayan teachings as the 'mayavirupa'.

182. Villard, 2011.

183. Gamma Rays, n.d..

184. Vidal, 2010.

Rainbow Body as a Technology of Consciousness

In the Dzogchen lineages of both the Buddhist and Bon traditions of Tibet, there are descriptions of advanced practitioners who, upon bodily death, transform the physical elements of the body literally into light. This phenomena, known in Tibetan as *jalus*, is often referred to in English as the cultivation of *rainbow body* due to the sudden appearance of rainbows in the vicinity of the master's body at the moment of and immediately following death. Upon attainment of the rainbow body, it is said that these beings are no longer restricted by time and space, and can appear and disappear on the physical plane *at will*. Here is the Dzogchen Master, Chogyal Namkai Norbu:

> When a person has that realization of Rainbow Body then their physical body slowly disappears and other people cannot see it. It seems as if that person has disappeared, but in the real sense he/she is alive and continuing their activities actively in the Rainbow Body. They can continue doing benefit, being active in the Rainbow Body for centuries and centuries, just like Guru Padmasambhava.[185]

In the occurrences when a master attains rainbow body it is reported that his or her physical body dissolves (or shrinks), sometimes leaving no trace of its once physical existence. Tenzin Wangyal Rinpoche, a lineage master in the Tibetan Bon tradition, writes of it in this way: "The realized Dzogchen practitioner, no longer deluded by apparent substantiality or dualism such as mind and matter, releases the energy of the elements that compose the physical body at the time of death."[186]

It is said that the phenomenon occurs for two reasons. First, it arises as a result of the fact that the practitioner's stabilization of Awakened Awareness as the base of all Reality has matured so fully that it begins to affect the physical body. According to this perspective, the experiential knowing of Absolute Reality is so well-established that it penetrates all the way down to the level of the atoms and cells. In this light, the attainment of rainbow body is also a sign of the full cultivation of the buddha bodies (Nirmanakaya, Sambhogakaya and Dharmakaya). At this level of realization, all bodies (physical, emotional, mental, soul, monadic) arise and are known directly as an expression of Awakened

---

185. Transcription of an oral teaching in Barcelona, Spain. Oct 3, 2010, by Chogyal Namkhai Norbu

186. Wangyal, 2002, p 141.

Awareness. Even the individual's karma has been transmuted and released.

Secondly, it is said that the rainbow body phenomenon occurs as an expression of compassion for the teacher's students and all those around him or her. To witness the phenomenon provides deeper evidence for faith to the students and helps to signify that the spiritual path is worth following. Rainbow body is the ultimate proof that a student needs to leave no doubt in his or her practice.

Over the past century, there have been several well-documented cases of both men and women who have attained rainbow body. The list of masters includes Shardza Tashi Gyalten (1935), Kenchen Tsewang Rigdzin (who reportedly transformed into rainbow body and disappeared alive in 1958), Ayu Khandro (1953), and Khenpo A-chos (1998) just to name a few of the more recent accounts within the Tibetan tradition. One of the most interesting reports that we receive from the students of these masters is that the teacher is often said to reappear, reconstituting him or herself in the form of a light body to offer final teachings. The founding of the Dzogchen lineage itself is said to have been initiated in this way. According to the tradition, the Master Garab Dorje returned in a body of light after his death to give his student Manjusurimitra a final set of instructions on how to perfect the Dzogchen view of ever-present radical awakening.

The transference of a fully awakened human being into a body of pure light additionally has a basis in many other traditions, such as the Tamil Siddha tradition, Taoism, Christianity, and many other lineages. According to the Trans-Himalayan understanding, the attainment of the rainbow body corresponds the passing of an advanced initiate through the 4th initiation. This involves the individual's full stability in radical awakening to Primordial Absolute Presence; stabilized identification in the buddhic altitude of the soul whilst opening more and more fully to the monadic; full transition out of humanity and into the planetary culture of Hierarchy; plane access increasingly opened to all seven sub-frequencies of the kosmic physical plane, and the being's transcendence of mind so as to operate fully through the intuitive mind wisdom-intelligence of the buddhic plane. Crucially for our considerations here also, it is understood that, subsequent to the 4th initiation, the initiate gains the capacity, should they choose it, to manifest on the physical-etheric, astral and mental planes through a mayavirupa, or body of light. Here is Djwhal Khul:

> He [the initiate] can work through a physical body (with its subtler sheaths) or not, as he sees fit. He realizes that he, as an individual, no longer needs a physical body or an astral consciousness, and that the mind is only a *service instrument*. The body in which he now functions is a body

of light which has its own type of substance. The Master, however, can build a body through which He can approach His incoming disciples and those who have not taken the higher initiations; He will normally build this body in semblance of the human form, doing so instantaneously and by an act of the will, when required. The majority of the Masters who are definitely working with humanity either preserve the old body in which They took the fifth initiation or else They build the "mayavirupa" or body of maya, of physical substance. This body will appear in the original form in which They took initiation. This I personally did in reference to the first case; i.e., preserving the body in which I took initiation. This the Master K.H. did in creating a body which was made in the form in which He took the fifth initiation.[187]

According to the Christian symbolism, the 4th initiation corresponds to the crucifixion, and the phenomenon of the rainbow body, or body of light, has intrigued Christian scholars and practitioners as trans-lineage dialogue between Eastern and Western traditions has opened up. Father Francis Tiso, associate pastor at Our Lady of Mt. Carmel in Mill Valley, and Brother David Steindl-Rast, a Benedictine monk, together investigated the potential connection between the rainbow body phenomenon and the resurrection of Jesus. As many of us will be aware, Jesus is said to have died and been placed in a tomb, only later for others to find that his physical body had vanished. Several different places in the Gospels mention that Jesus reappeared to offer teachings, blessings, and insight to his disciples.

A few examples of Jesus reappearing are helpful to land this point. Days after his death, Jesus reappears to eleven of his disciples on a mountain in Galilee. The disciples became frightened and thought he was a ghost. After assuring them that he was as real as they were by joining them for a meal, he offered them a teaching and direct transmission:

> 'This is what I told you while I was still with you: Everything must be fulfilled that is written about me in the Law of Moses, the Prophets and the Psalms.' Then he opened their minds so they could understand the Scriptures. He told them, 'This is what is written: The Christ will suffer and rise from the dead on the third day.'[188]

---

187. Bailey, 1960, p. 705.

188. Luke 24:44-46.

Although the apparent physicality of his being is quite real, it is clear that subsequent to the crucifixion, Jesus is not bound to physical reality. He seems to appear and disappear at will. Other cases are present throughout the New Testament: on the road to Emmaus he appears to Cleopas and one other disciple; he appears to Mary Magdalene and another Mary near his tomb; and to several others near Lake Tiberias.

Interestingly, the research conducted by Father Tiso and Brother Steindl-Rast reported that all those individuals who fully stabilized Awakened Awareness and attained rainbow body also exhibited an extraordinary capacity for both compassion and moral action.[189] This implies, from our perspective, that enlightened behavior and conduct is a key piece of the puzzle for manifesting rainbow body (from an integral perspective, this would mean the full manifestation of awakening in all Four Quadrants). If realization is to penetrate right through to one's cells it requires that the whole of one's life becomes a vehicle for profound compassion and integrity in the world. Indeed, we would expect nothing less from the level of evolutionary advancement that marks the 4th initiation.

Kosmic Travel without a Trace

Returning to our consideration of the possibility of other humanities already having some presence on Earth, we first note that it is our understanding that the cultivation of rainbow body seems to be an advanced evolutionary potential for humans here on Earth for which we already have some evidence. With this being the case, we speculate that these same capacities for radical and evolutionary awakening, or corresponding forms of them, may exist on exoplanets beyond our Earth, where extraterrestrial civilizations could have developed. This speculation that other civilizations may also have the capacity for rainbow body naturally rests on the understanding that the possibility for radical awakening to Awakened Awareness is a *kosmic constant*. That is, radical awakening is available to all civilizations, on all planes, and on all planets, to the degree to which Awakened Awareness is operating through forms sophisticated enough for it to Self-Recognize. Indeed, such speculations already have a base in the tantras of the Dzogchen tradition. The *sGra thal-'gyur* Tantra, for instance, suggests that Dzogchen teachings on radical awakening to the Natural State of Absolute Reality

---

189. See Tiso's "The Rainbow Body Phenomena" audio lecture sponsored by the Institute of Noetic Sciences for more details. http://noetic.org/library/audio-lectures/the-rainbow-body-phenomenon-with-father-francis-ti/

"are found in many different inhabited world-systems throughout the great universe", and that while it is traditionally understood that there are some 6,400,000 extant within Dzogchen, only a comparatively small number of these exist on Earth.[190] In the same text, as translated by John Reynolds, we read that the "Dzogchen Tantras hold particular importance in thirteen great star systems (one of which is our own)".[191]

For those civilizations who are more advanced than us, and with the above point in mind that some civilizations might have even had up to 1.8 gigayears longer than us to get there, it is not far-fetched to speculate that the rainbow body phenomenon might be something that *all beings* of a particular extraterrestrial planetary culture could have access to. At such a point in spiritual advancement, an entire culture could no longer be properly understood as physical plane beings as humanity exists on Earth, but rather as inter-dimensional beings consciously inhabiting multiple planes. Continuing the train of thought further, we speculate that those advanced kosmic cultures with the capacity for rainbow body may have the ability to appear and disappear at will throughout multiple planes. For these cultures, such a capacity would be a standard form of consciousness-technology. Indeed, this set of suppositions matches very well the first person accounts of some human beings who claim to have had ongoing contact with extraordinary beings seemingly unknown on Earth.

The Kosmic Federation

In the Trans-Himalayan transmissions, it is understood that there are already beings of extraterrestrial origin and colossal levels of evolutionary advancement here on Earth, working within the soul and monadic altitude planetary cultures of Hierarchy and Shamballa. This relates to our ability to understand the kosmic paths spoken about earlier, and particularly the first of them, not just as paths leading out and away from the Earth but also as paths of energy *leading in*. What is also understood in these transmissions is that as our collective evolution unfolds, humanity may become aware not just of great individual beings whose origin may be extra-planetary, but of the other kosmic civilizations and cultures we have been speaking about herein. This builds upon the point made by Sir Martin Rees earlier that it may not be until we have collectively begun to unfold higher evolutionary

---

190. Reynolds. 1996, p. 13.

191. Reynolds. 1996, p. 13.

potentials and capabilities that we can detect their existence. Here again is Djwhal Khul:

> The evolution of consciousness and of matter require each other. At their root, both consciousness and matter are the One Life which does not evolve. As souls mature in consciousness they become aware of other souls who have also matured to that level. So it is with civilisations. We will become aware of other civilisations that exist or have existed or will exist in our galaxy as well as the wider universe, once we have evolved to the level upon which that consciousness is operating. Once an individual awakens as a soul it is also able to perceive how it has already been acting throughout the time and space of many lives. Once a planetary soul or the soul of a civilisation awakens it becomes aware of what it has been trying to achieve, and which other souls have played a part in that process. We do not suddenly open to receiving input from the Sirian system for example; we awaken to the fact that we are already part of (and have always been a part of) that system—in fact a seed of it. There is no longer a civilisation operating on a planet orbiting one of the Sirian suns. It evolved to a point where it was able to seed a whole region of space/time. An example on Earth would be the Tibetan culture which is dying out in its homeland but spreading the seeds of its wisdom throughout the planet, blending them with the other cultures that are already present.[192]

This is an extraordinary point—one that calls us to expand dramatically our collective understanding of ourselves from a worldcentric perspective to one that is kosmocentric. It makes sense to us that we will not be welcomed fully into the kosmic community until we can function as a planetary "We". A planetary "We" would require us to speak as one voice, not simply from the position of a parliament or group of representatives in a world federation but rather as a planetary council of beings so developed, so awake, and so in-tune with the human experience that they can speak kosmically as *Earth humanity*.

Furthermore, what the above quotation also suggests is that as humanity evolves, one thread of recognition that we can expect to emerge is our collective realization that *we ourselves are a kosmic culture and civilization already*—one deeply connected to and an expression of an already extant collective culture and civilization of the kosmos, and

---

192. Lyon, 2010, p. 76.

that we are here to serve as such. The consideration of humanity not just having a service role within the multidimensional life spheres of the Earth, but outside of the Earth as well, points to entirely new dimensions of our collective evolutionary potential. Once more, Djwhal Khul:

> As an initiated consciousness begins to form the basis for the coming civilisation on Earth it will coincide not only with the great revelation of essential divinity but also with the growing awareness and contact with other 'human' civilisations within our galaxy. There will be those whom we are in a position to help and others who are in a position to help us.[193]

This leads us to a profound place of contemplating humanity as a *kosmic server*—a kosmic disciple, both learning from those more evolved civilizations of the kosmos and contributing to those who are younger. It calls us to begin to envision Earth, as it may exist in the future, as not just containing many multidimensional cultures, kingdoms, and their various energetic or physical fields in its LL and LR quadrants, but as one constituent holon among many others in the LL and LR collective fields of the kosmos. To the degree that we are able to envision this, we begin to see the role of humanity as a single kosmic server with billions of unique faces, working with brother and sister cultures and civilizations of the kosmos in a Universal Federation of Light so as to actualize universal Purpose.

All of this requires and entails Earth itself coming to reveal and express its own unique planetary Purpose. It is from this great universal context, which we have explored in the last four chapters, to a specific consideration of the specific role it is the Purpose of the Earth to play, that we now proceed.

---

193. Lyon, 2010, p. 315.

Chapter 20 - Kosmic Community

Part 7: Sacred Earth

The Life principle originates from the eternal and infinite realms and yet it expresses in the worlds of becoming as the evolutionary impulse. Initiates are the carriers on Earth of that Life, and Earth itself is a seed of that Life in its galactic environment.

(Master Djwhal Khul and Bruce Lyon)[194]

---

194. Lyon, 2010, p. 194.

Earth is Eden

Chapter 21
The Purpose of Earth
Jon

In the previous sections we have explored radical awakening, evolutionary awakening, post-mastery stages of unfoldment, and the multidimensional holarchy of the kosmos. In Chapter 19 we started to explore the unique Purpose of the Earth in the kosmos, and in Chapter 20 we explored the great space of kosmic community within which this is potentially unfolding. Now, we will explore in detail the emerging revelation of the sacred Purpose of Earth in the kosmos, as it is being progressively revealed in the Trans-Himalayan teachings.

In his most recent body of teaching through Bruce Lyon, Djwhal Khul gave what he called a "sanctioned hierarchical revelation"[195] of that aspect of the kosmic Purpose envisioned and transmitted by the galactic Logos of our Milky Way galaxy, for which the planetary Logos of the Earth has taken responsibility. The extent to which he unpacked this teaching was unprecedented. Only once or twice before had the teaching gone so deeply into that sacred Purpose for which it is understood the Earth exists, and which will only begin to reveal itself once love is established as the basis of humanity's relationship not just with itself, but with all other evolutionary life-spheres on Earth. One of those times was in his transmission through Alice Bailey, where he gave the following four points:

1. The first aim and the primary aim is to establish, through the medium of humanity, an outpost of the Consciousness of God in the solar system. This is a correspondence, macrocosmically understood, of the relationship existing between a Master and His group of disciples. This, if

---

195. Lyon, 2010, p. 291.

pondered on, may serve as a clue to the significance of our planetary work.

2. To found upon earth (as has already been indicated) a powerhouse of such potency and a focal point of such energy that humanity—as a whole—can be a factor in the solar system, bringing about changes and events of a unique nature in the planetary life and lives (and therefore in the system itself) and inducing an interstellar activity.

3. To develop a station of light, through the medium of the fourth kingdom in nature [humanity], which will serve not only the planet, and not only our particular solar system, but the seven systems of which ours is one. This question of light, bound up as it is with the colours of the seven rays, is as yet an embryo science, and it would be useless for us to enlarge upon it here.

4. To set up a magnetic centre in the universe, in which the human kingdom and the kingdom of souls [Hierarchy —the planetary meta-sangha] will, united or at-oned, be the point of most intense power, and which will serve the developed Lives within the radius of the radiance of the *One About Whom Naught May Be Said* [the galactic Logos incarnating through our entire Milky Way galaxy].[196]

Kind of amazing, no?! In this most recent transmission with Bruce Lyon, Djwhal Khul took it into even greater detail. From *Occult Cosmology*:

When Earth humanity looks into the universe one of its deepest questions concerns whether or not there is other life out there. Worded another way this question might be "Are we alone here?"

What remains as yet hidden to the human psyche is that humanity has the potential to be a source of life in our cosmic environment. We are that life arriving. This needs to be explained because I am not talking here about seeding biological life or at least not directly.

When we look out into the cosmos we are looking along a particular wavelength or frequency—like being tuned to a radio station or operating only in ultraviolet. Earth is a planet of deep material manifestation. Of

---

196. Bailey, 1942, p. 217.

course the universe is alive with consciousness on etheric levels and higher, but there is little conscious intelligent life at the depth of prakritic immersion experienced by Earth humanity. So we are both alone and not alone....

Our Logos, being a rebel and willful but also stimulated by love as a light-bringer, chose to go beyond the current limits of the extension of unified consciousness in the galaxy. Whether the experience of Earth will serve as a warning or an encouragement remains as yet undecided but the potential is for Earth to help the evolutionary advance of life in the galaxy in a potent way....

Our Logos is part of the cosmic Dweller [that part of the consciousness of the galactic Logos that is below the threshold of its consciousness] and when elements of the subconscious rise and join consciousness the range of that consciousness is expanded. If enough human souls can regain conscious polarisation on the buddhic plane and take their place in the ashrams, the collective being called humanity can begin to radiate the essence of its experience into the self-consciousness of the cosmic system....

The buddhic plane is part of the cosmic ethers and Earth humanity can serve as a vitalising radiatory power that carries with it the experience of having being deeply immersed in the material world and returned.

This vitalising power can help other civilisations which exist on more subtle levels to manifest—to come more fully IN to their vehicles.[197]

As we said in Chapter 19, for those buddhas who have graduated from their planetary school so as to become Logoi, the kosmos is a field of much greater freedom in evolutionary experimentation. From their radical wakefulness and identification with kosmic Purpose as it is envisioned and transmitted by the kosmic Logoi within whose bodies they find their place, they are able to take responsibility for the unfoldment of a unique expression of it through their own fields of manifestation. What Djwhal Khul is exploring in the quotations above is the idea that in choosing to ensoul such a dense field of planetary manifestation as the Earth, our planetary Logos actually went against the kosmic grain, as it were. This relates to the mystic appellations by which this being is known to those liberated masters and buddhas of Hierarchy and Shamballa: "The Divine Rebel" and "The Great

---

197. Lyon, 2010, p. 291-292.

Sacrifice." This is related also to his statement that while the existence of evolutionary kingdoms of life beyond the Earth is by no means rare on planes subtler than the physical, self-conscious biological life on the physical plane *is* rare.

What Djwhal Khul is also saying above is that the purpose for which this being made this choice goes further than the evolution of the biological forms necessary to support and express a planetary culture's recognition of Absolute Source. It is something more specific; something dependent on the sacred "vitalizing power" present in deepest matter.

This leads us nicely into the mysteries and revealed divinity of the physical body, and the mighty service it can offer once radical awakening and the shift of evolutionary awakening starts to shift into the monadic. As we explored in Chapter 10, when this process begins to unfold there are profound corresponding changes in the body. As was discussed there, the physical body and the will-to-live held in the base chakra are brought into contact with and known as one with the monadic Life in the heights:

> As this unfolds, the physical body is transfigured into a temple of Eden, and the fusion of radical wakefulness, monadic Life and physical embodiment is grounded so fully that the cells of the body begin to literally shine with light, as has been demonstrated by some of the greatest Masters and teachers of the past. This allows the monadic Life-essence at the core of every cell and atom of the physical body, the product of Absolute Reality and its Life-force expression's apparent self-contraction into the form of a personal bodymind, to become activated so as to saturate every muscle, nerve fibre, bone, and organ of the body in unbroken bliss. And it allows the black hole capacities of the monadic bodymind to bend and warp space and time in conformity to divine Purpose, to become the property of a Master on the physical plane. Then, the powerhouse of their very physical presence becomes an innately purifying, healing, empowering, and stimulating force within their environment, in an incredibly potent way.

The kosmic Purpose of the Earth is found in the unfoldment of this process in relation to the planet as a whole. The use of the word 'Eden' in the title of this book is in reference to the unconditional divinity of all matter that is realized when a) radical awakening is extended all the way into the Nondual so that the entire Gross kosmos is recognized as an infinite matrix of expression for, and not differing from, the

Primordial Presence of Awakened Awareness; and b) evolutionary awakening moves into the monadic altitudes. Then, not only is all form revealed as unconditioned Godhead, but it is also revealed that matter, at its heart, has *power*.

This power is the presence of kundalini in the base chakra and the body as a whole. As described in Chapter 10, kundalini is the anchored presence of the monadic Life in the base chakra. It is the sub-conscious psychological force of the will-to-live. It is held there by our self-contraction and misidentification with form, which prevents both our naked opening to the Awakened Awareness that we are and our immediate identification as its kosmic Source-Energy. When that self-contraction and misidentification begins to be released, however, this trapped energy is freed. It begins to rise and circulate through our system according to nature and divine intent of Awakeness itself. As is the testament of many of the great realizers, once this occurs our entire system of embodiment—mental, astral, etheric, and physical—becomes emphatically rewired and reconfigured in harmony with awakening. The radical Wakefulness and monadic Life at the core of every cell is awakened, and the body starts literally to *shine*.

Sacred to understanding this process at a planetary, solar and galactic scale is the Trans-Himalayan teaching that the Earth embodies the base chakra of the solar Logos incarnating through our solar system, and is thus a reservoir of *kosmic* kundalini and the *galactic* will-to-live.

Exploring the specifics of this requires us to go deeper into the wisdom held in the Trans-Himalayan teaching on the holarchical fractal scaling existing between beings incarnating at different octaves (i.e. intra-planetary, such as a human being, and planetary, such as a planetary Logos). As was described in Chapter 9, just as the kosmic physical plane has seven sub-planes, so also does each of these sub-planes have seven sub-frequencies of increasingly subtle energy-matter. The sub-frequencies of the densest of these, the physical plane, are divided into two main forms of vibrational substance—dense physical and etheric. The lowest three sub-planes are those of dense physical matter—gaseous, liquid and solid—whilst the four higher sub-planes are etheric. Therefore, a human or an animal's physical body, solid, liquid and gaseous, is understood to be composed of matter of these three densest sub-planes of the physical, whilst their etheric body and all etheric bodies (those of planets, animals, and humans) are understood to be composed of the etheric energy-matter of the higher four sub-planes. This can be seen in Figure 23. Notice how for a human monadic essence in physical incarnation, the lowest three sub-planes of the kosmic physical plane provide the dense, liquid and gaseous substance for their physical body. Now notice how for a kosmic Logos in physical incarnation through all the shown kosmic planes, what are the physical, astral and mental planes for a human being provide the dense, liquid and gaseous substance for its physical body.

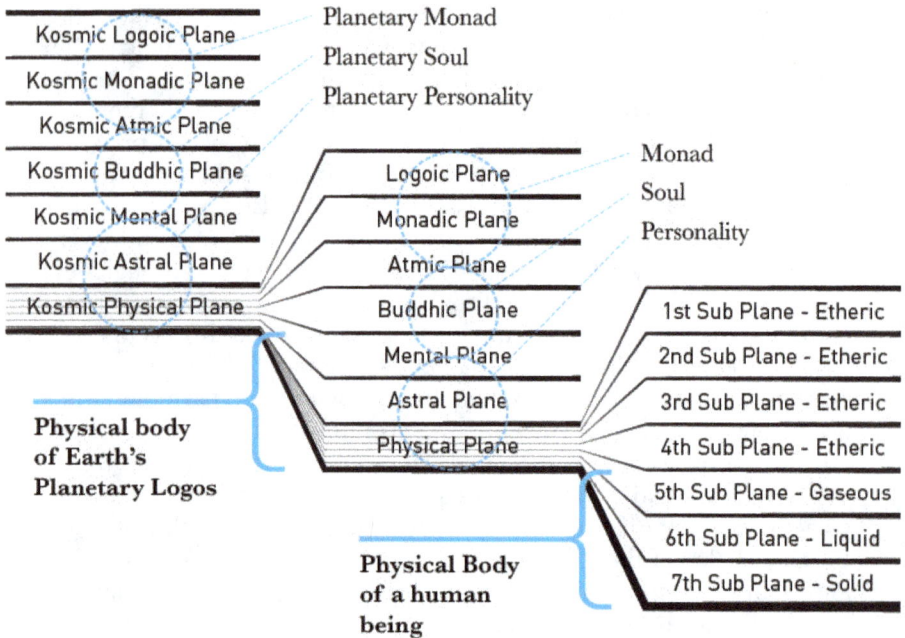

Kosmic Logoic Plane		Planetary Monad
Kosmic Monadic Plane		Planetary Soul
Kosmic Atmic Plane		Planetary Personality

Kosmic Buddhic Plane — Logoic Plane — Monad
Kosmic Mental Plane — Monadic Plane — Soul
Kosmic Astral Plane — Atmic Plane — Personality
Kosmic Physical Plane — Buddhic Plane — 1st Sub Plane - Etheric
— Mental Plane — 2nd Sub Plane - Etheric
Physical body of Earth's Planetary Logos — Astral Plane — 3rd Sub Plane - Etheric
— Physical Plane — 4th Sub Plane - Etheric
Physical Body of a human being — 5th Sub Plane - Gaseous
6th Sub Plane - Liquid
7th Sub Plane - Solid

Figure 23. Chart showing the fractal nature of the planes as taken from *A Treatise on Cosmic Fire*.[198]

Now, when we come to consider a being incarnating at a higher octave, such as the planetary Logos of the Earth, all of this is scaled up fractally.[199] Just as the three densest sub-planes on the physical plane are those of our own gaseous, liquid and solid embodiment, with the subtlest four sub-levels being those upon which our etheric bodies

---

198. Bailey, 1925, p. 344.

199. In the Trans-Himalayan teaching all octaves of incarnation higher than the intra-planetary up to but not including the universal only involve a fractal up-scale by a factor of one. That is, all beings incarnating at galactic, super-constellational, constellational, solar, and planetary octaves are understood to have their main centre of gravity on the seven kosmic planes (kosmic physical, kosmic astral, kosmic mental, kosmic buddhic, kosmic atmic, kosmic monadic, kosmic logoic). When we come to speak of beings incarnating at the universal octave and beyond, we then scale up fractally again. This would mean that all seven kosmic planes would now form the seven sub-frequencies of the *universal* physical plane.

reside, when we come to speak of a being incarnating at the planetary octave we shift everything up in scale. Then, the three densest sub-planes of *the whole kosmic physical plane* form the levels of gaseous, liquid, and solid physical embodiment, with the subtlest four sub-levels being those upon which the planetary etheric body is found.

From this perspective, fascinatingly, the noosphere of all human thought as it exists on the mental plane constitutes the gaseous or electro-chemical transmission field of the planetary Logos' dense physical body and central nervous system. The sphere of all astral relation embodies its liquid, circulatory aspect; the sphere of all dense physical matter on the planet its solid flesh. This demonstrates the relative nature of that which is considered dense and subtle energy-matter depending on the perspective of the one viewing it. From the intra-planetary octave of incarnation, such as might be viewed by a human being, the etheric, astral and mental planes are constituted of subtle energy-matter, non-perceptible to the five senses. From the perspective of the planetary Logos of the Earth incarnating at the planetary octave, they are all physical.

The present state of things on the mental, astral, etheric and physical planes of the Earth—the prevalence of war, poverty, world hunger, the exploitation of human beings and the Earth by human beings, the corruption and lack of transparency in virtually all human systems—is an expression of the lack of continuity of consciousness established by the planetary Logos as yet between its monadic, soul and personality depths of identification. That continuity extends presently from the planetary Logos's soul bodymind, down through its mental and emotional bodies, and into its etheric body, where the planetary cultures of Shamballa and Hierarchy compose its crown and heart chakras respectively. Not yet has the continuity been extended down into the physical body of the planetary Logos from planetary soul levels. And yet increasingly consciousness of the kosmic Love at the heart of the planetary soul, and the multidimensional nature of the miracle that is the Earth-being, is being translated into pockets of culture and social systems.

Crucial to this is the phenomenon Djwhal Khul is speaking of when he describes the kosmic service role that may come online for humanity once enough human beings shift their altitude onto buddhic levels. As a result of this they transition in their center of gravity from the physical into the etheric body of the planetary Logos of the Earth. From tht level of altitude they can then, as LL cultures and LR social systems, begin to naturally radiate the energy of their liberation into these subtle and populated levels of the kosmo. Additionally, through the continuity of consciousness that they, as LL cultures expressed in LR social systems, are able to establish between soul and personality embodiment, they are able to serve the planetary Logos's extended continuity of planetary consciousness *into* its physical body, which may then be revealed and empowered as Nondual Eden.

Both of these points relate to the role of the Earth as the solar base chakra and human beings as galactic kundalini. Here is Djwhal Khul:

> A human monad is essentially a spark of the one flame—the One Self. Its confinement on the lowest of the cosmic planes is analogous to the confinement of the kundalini force within the base centre, and is why our prototype, [the planetary Logos of the Earth], is known as the Great Sacrifice.
>
> As the human monad begins to realise itself as the planetary life, the solar life, the galactic life, the universal Life, so the highest and the lowest on the cosmic planes meet and cosmic kundalini rises. Human monads are therefore the energy of spirit-as-Purusha, as Atman, as Brahman sleeping within the cosmic physical plane. The triplicity of monad-soul-personality are the three serpents behind whom stands the Great Dragon who in his essential nature is the One Life.[200]

This is a teaching that when extrapolated to its kosmic correspondence reveals the top-down perspective on the kosmic paths. When we consider the kosmic paths from a bottom-up perspective, we envision masters engaging wondrous trails of unfolding into extraordinary new horizons. We see them extending their awakening to the Nondual Great Perfection to unfathomable new scales, and their depth of identification, relationship, plane access and kosmic intelligences into ever greater and more expansive spheres.

If we explore this phenomenon from a top-down perspective, however, we see that as human beings enter into radical awakening and are liberated through their shift in altitude from personality to soul to monadic levels, they rise as *units of galactic kundalini* along the kosmic nadi channels of the subtle energy bodies of planetary, solar and galactic Logoi. In this, the energy of kosmic self-contraction within the planetary body is released; further voices are added to the roar of victory that ever sounds on the kosmic planes, honoring the kosmic bodhisattva activity of the planetary Logos of the Earth; and more of the planetary body is *opened* to melt in recognition of infinite Awakened Awareness. Or, among those who take the first kosmic path, that of "Earth Service", and who choose to stay, we see units of galactic kundalini remaining within the aura of the solar base chakra. There,

---

200. Lyon, 2007, p. 121-122.

they ground—like a sine wave of awakening rippling through the field of kosmic matter—the naked Reality of Awakened Awareness in the very nuclei of our sentient, biological cells, in the molecules they are composed of, and in the soil of the Earth. Again, Djwhal Khul:

> *It is possible for kundalini to awaken but not arise.* We might liken Shamballa to the dark pupil of an eye. It is possible for the pupil to dilate or contract in size to allow more or less light or consciousness through. The degree of dilation is controlled by the number of Sixth Degree initiates who are asked to remain in Earth service. By refraining from rising, the centralised Life energy empowers and vitalises the ring-pass- not in which it remains willfully limited.[201]

To the degree that we together make it through the present global crises that face us so as to ground this Love fully as the basis of world culture and civilization, the Trans-Himalayan teachings understand the potential future for Earth in the kosmos to be profound. Ultimately, both of these processes—the choice of more and more liberated human beings as galactic kundalini to rise or to stay—will be key. The deepening of the planetary Logos of the Earth's integration of kosmic radical awakening and evolutionary awakening into its kosmic monadic altitudes will, equally occupy a fundamental position. It is this that will allow the release *and* intensification of the vitalizing power of the will-to-live of which Earth is a galactic reservoir. It will allow the monadic Life at the heart of each and every particle of energy-matter composing the planetary body to be awakened into luminous flame. It will evoke the revelation of the whole planetary body—the geosphere, biosphere and noosphere—as Eden, and the calling forth of those civilizations, presently abiding on subtle levels, into physical evolutionary emergence.

Any one of us can look out to the natural world right now and be utterly awed and humbled by its breathtaking beauty. It is equally awe-inspiring and breathtaking to imagine what more beauty may be revealed on this sacred planet as the forms of the natural kingdoms develop and unfold so as express the kosmic love and wisdom of the planetary soul, and then the invincible Life of the kosmic monad.

It is as this unfolds that humanity may enter into that sphere of kosmic service that Djwhal Khul speaks of in the fourth point quoted at the beginning of this chapter. There, he prophesizes a future time when Earth resides as a magnetic center in the universe in which humanity, fused so fully with Hierarchy that full continuity of consciousness exists throughout all levels of the multidimensional being that is the planetary

---

201. Lyon, 2007, p. 122.

Logos of the Earth, exists as one of the most powerful points of radiatory service in the galaxy. As we explored at the end of the end of Chapter 18, by 'humanity' here we mean not just Earth humanity, but humanity as that evolutionary life sphere in which self-consciousness emerges and allows the emergence of full radical and evolutionary awakening, as it expresses throughout the universe. As the solar base chakra, Earth serves as a reservoir of the kosmic will-to-live. As this kundalini force is activated, it radiates from our planet out into the space of the kosmos to be drawn upon by other civilizations, such as we discussed in Chapter 20, residing as yet only on subtle planes. In this way, the awakening of humanity as kosmic kundalini contributes to the continued full incarnation and Self-recognition of universal Spirit, and the emergence of a physically embodied kosmic Federation of Light.

Chapter 22
A Planetary Lineage of Reality
Jon

The last chapter explored a grand vision for the potential future of Earth. As we together continue our evolutionary journey forward from the present, what steps do we see as most relevant to the safe passage of our crisis-ridden planet ever deeper into the unfoldment of this sacred kosmic potential?

Without doubt, the transformation of our global political, religious, economic, cultural, social and environmental modes of engagement is paramount. At present, humanity, the natural kingdoms of the Earth, and the Earth itself have little effective representation. Multinational corporations do. The military-industrial complex does. The financial interests of nation states and the wealthy (but comparatively few) do. Humanity as a whole, the natural systems of the Earth and the Earth itself, however, do not. The current paradigm, rooted in the dominant ethnocentric and individualistic worldviews, has led our planet to a crisis point of greater scale than we have ever before faced. Now, as we reconcile ourselves to such imminent possibilities as runaway climate change, a potential global water crisis and the global energy crisis (the possibility of peak oil, for instance), we together stand before a predicament so large that our current approach, rooted as it is in separation, personal ego and survival fears, does not have the answer.

Never before have our collective fates been so intimately interwoven with each other. We stand before an initiatory burning ground of such magnitude that it must *necessarily* call forth from humanity a whole new form of engagement with itself, and the Earth: an entirely new form of culture and society rooted in the ethic of Freedom and Love that is revealed when radical awakening has been recognized and identification has begun to open to the soul.

329

This freedom and this love are both unconditional. Neither depends on our collective survival of the challenges we currently face; hence, when considering how we collectively journey from here, Djwhal Khul and Bruce Lyon suggest:

> Let us step back into the universal soul. In the hundreds of billions of galaxies, each with hundreds of billions of stars, the universal soul has been experimenting with biological life and consciousness over time scales that are difficult to grasp. Civilisations come and go in kosmos just as they do in our galaxy, in our solar system and on our planet. Each civilisation is harvested by the universal soul; which is why it is possible for evolution to accelerate. What once took millions of years can now be accomplished in hundreds of thousands of years, then thousands and then hundreds and then tens of years.
>
> The current civilisation on Earth will eventually end just as an individual life ends. What is important is not the survival of the civilisation but its harvest. *Therefore it is more important for human consciousness on Earth to evolve than it is for human life to survive.* Think about this, for it is a key statement that underlies the crisis that humanity is passing through. As the resources on planet Earth begin to be used up, as the climate changes, as population pressures grow, will humanity continue to evolve its consciousness into an expansive love, or will it begin to contract and spend all its evolutionary potential on trying to prolong physical survival at all costs?[202]

And yet as this is written, new forms of consciousness, behavior, culture and society, rooted in the soul-ethic of Love, the recognized One and deeper structures of wisdom-intelligence *are* emerging. From new forms of political engagement that are genuinely seeking to serve and represent all beings and the Earth; new forms of interest-free banking that seek to empower prosperity and economic freedom without enslaving us in ever-mounting debt; efforts to share the world's abundant resources amongst all beings, regardless of race, creed, gender, or financial status; the growing interface between neuroscience and mystical spirituality that is seeking to understand the UR referents to the states of UL awakening cultivated by spiritual practitioners and their lineages; and new forms of media and technology-savvy art that are celebrating a new renaissance of Light in human creativity, ever stronger tides of integration and healing are rising.

---

202. Lyon, 2010, p. 73-74.

Empowering this rising wave of change are the energies being transmitted into the noosphere, biosphere and geosphere by the planetary cultures of Hierarchy and Shamballa. In the Trans-Himalayan teachings it is understood that these two centers constitute the heart and crown chakras of the planetary Logos of the Earth, respectively. Furthermore, in these teachings it is taught that at the present time of planetary evolution, owing to the galactic current of empowerment pouring into our solar system and Earth as a result of the solar and planetary Logoi deepening identification with and as the emanated Life of the galactic Logos,[203] there is an activation of the planetary chakras occurring. And now, not just the kosmic love-wisdom transmitted by Hierarchy, but the kosmic Will, Purpose and monadic Life of the planetary Logos of the Earth, held in Shamballa, are being released into humanity *for the first time*.

Unpacking this, prior to the Second World War, it is understood that this Shamballa transmission was only permitted release into the planetary meta-sangha of Hierarchy, where the kosmic Will and Purpose of our planet was translated into the evolutionary Plan for that particular cycle. This was owing to decision of the enlightened bodhisattvas and sages of the Earth that humanity and the other life-spheres of evolution were not yet ready to respond to its potency without destructive effects. The impact of Shamballa's transmission in a planetary sense, and contact with monadic transmission in an individual sense, is understood to have three effects: purification, destruction and organization.

Purification relates to the consumption in the flames of radical Truth of our misidentification with ego and separation, as well as their lines of extension through our individual daily lives and collective systems of interaction. Destruction relates to the literal destruction and shattering of bodies and fields of energy-matter that may eventuate from the transmission of pure electric Life pouring in from Shamballa. Life is that principle of Being whose "unfettered positivity"[204] liberates consciousness from form. When consciousness is not sufficiently free, and clings onto the form, however, it is the destruction of the form that may result. Organization should be understood as Self-organization—the capacity of the dynamic Life-force of Source to re-organize itself naturally and spontaneously through individual and collective systems when purification and destruction have done their work.

---

203. A phenomenon the significance of which is out-pictured in the conjunction between the solstice sun and the galactic centre between 1998-2012.

204. Lyon, 2005, p. 304.

Fundamentally, the transmission of Shamballa is one of electric empowerment. It is understood that for the millions of years since its founding on Earth, Hierarchy has shielded the natural kingdoms and humanity from this transmission until such a time as humanity's consciousness was sufficiently evolved for Hierarchy's empowerment to vitalize the activity of unity and goodness, rather than egoic separation. Since the Second World War, however, which is understood to have faced humanity with such incredibly humbling crises that the walls of separation in its heart and mind were torn down and opened to a whole new level,[205] the cyclic release of this fierce transmission into humanity and the entire Earth system has been instituted to occur every quarter of a century. This began in 1975, then 2000 and then in 2025, and so on. In the Trans-Himalayan teachings, these are described as the "Shamballa Impacts".

As a result of the commencement of these Shamballa Impacts, and as can be seen from a glance at the world, the Fire of Truth and the Force of Awakening that flows from this center of planetary power has begun to impact humanity *globally*. This is occurring both within and outside of the traditions, cleaving through all obscurations to the Absolute—Causal, Subtle, and Gross—resulting in Nondual awakening breaking forth among individuals globally to an unprecedented scale, and a deeper and deeper capacity of human beings to identify as kosmic monadic Life. It is calling forth the opening up of multidimensional relationship with the communities of the subtle planes and penetration into new planes of experience and their energies. It is evoking the emergence of new structures of wisdom-centered intelligence; the emergence of new levels of biological complexity sufficient to support their incarnation on the physical plane; and an increase in earthquakes, tsunamis and volcanic eruptions as the inner fires of the body of the Earth are increasingly stimulated.

As was explored briefly in Chapter 2, it is understood within the Trans-Himalayan teachings that their role has been and is to hold a space of relationship. That relationship extends between the different traditions and fields of human activity, and between humanity and post-human kingdoms of life. As such, it is intended to support humanity to learn to respond collectively to the transmission of such post-human spheres and establish a continuity in consciousness between the various multidimensional levels of the planetary being and the kosmos. As we also noted, however, it is understood that once humanity increasingly *does* learn to respond to their transmission, and to open to recognition

---

205. According to Don Beck, the first integrative developmental structure, yellow systemic, began to emerge around the middle of the 20th Century. (http://www.en-lightennext.org/magazine/j22/beck.asp)

of its own radical Absolute and evolutionary multidimensional Self, the force of awakening pouring forth from Shamballa into humanity may increasingly wash away the mediating principle of the Trans-Himalayan movement itself.

It is this which will contribute to the breaking forth of the revelation of awakening on a global scale, which as it tetra-emerges and tetra-meshes from and through all Four Quadrants may contribute to the emergence of a Planetary Lineage of Reality on the physical plane. The Truth of Ultimate Being, or Awakened Awareness, has no lineage, and is the property of no particular culture or civilization on Earth or elsewhere. It is neither new nor old. It is simply the Truth of what is, what has always been and what always will be.

As we explored in Chapter 2, a Planetary Lineage of Reality has always been here on the subtle planes and the secret places of the Earth in the form of Hierarchy, to represent Ultimate Source. Over the course of humanity's evolution, Hierarchy has been living and transmitting these profound truths of Reality, the kosmos, and Earth; supporting human beings to move into deeper horizons of unfoldment and liberation, and yet owing to the stage of collective evolution on Earth, this lineage has remained esoteric and unseen—living its truths primarily on the subtle planes. When human beings *have* awakened to some of the very same realities that form the basis of culture and civilization for Hierarchy over the course of history, and have then translated that individual experience into collective cultures and societal systems, the River of Truth has been cloistered into ethnically- and nationally-situated lineages. As a result, it is often very difficult to see where the contours of that awakening end and where culture begins.[206] Indeed, this is the very point made by Integral Theory concerning the subjective, biological, cultural, and socio-historical relativity of all perspectives. No matter how deep our experience of Reality, our self-report and hermeneutic interpretation of it will always be colored by the depth of identification and structural intelligence from which it is enacted, the biological equipment through which it has occurred, the cultural milieu in which it is arising, and the historical circumstances of its occurrence.

Something that is occurring today, however, at an unprecedented speed, is the emergence more and more fully within humanity of altitudes of consciousness, biology, culture, and civilization that are no longer ethnocentric, but rather worldcentric, and moving into

---

206. The notes made in Chapter 20 on how the practice of Dzogchen is understood within that lineage to exist within many other star systems than our own, but which are now deeply embedded within the Nyingma Vajrayana and Bon religious traditions, are a case in point.

kosmocentric. This will increasingly provide the emergence of a global riverbed of culture and civilization whose grooves are deep enough for the Planetary Lineage of Reality that has always been here on subtle levels to pour into the embodied levels of human life. What this means is that as the Planetary Lineage of Reality externalizes into human culture and the soil of Earth, it will begin to enact the truths of the Four Quadrants at an entirely new level to that which Awakening has been expressed through before. With the Fire of Truth raining down with increasing impact upon human awareness; the in-pouring force from Shamballa of pure monadic Life; the opening up of contact between humanity and other communities of lives and other planes; the unfoldment of entirely new forms of intelligence, and the biological levels of evolutionary complexity necessary to support all of these, we perceive that extraordinary changes are occurring on Earth. Indeed, we envision that one day awakening itself may be released from being an experience particular to the mountain retreats, monasteries, ashrams and hermitages in which it has been so carefully safeguarded, and gratefully so, to be taken on as the birthright and natural inheritance of all humanity. As the continuing emergence of awakened realizers who have no affiliation with any particular tradition testifies, this has already begun.

What is coming, should we find the ethic of love within ourselves quickly enough to navigate the precipice we currently stand on, is a kosmocentric culture and civilization of Reality—a planetary civilization grounded in the Unconditioned Absolute Presence arising as the entire evolutionary spectrum of the kosmos in all its majesty; humanity as an awakened Earth Community, self-recognized as a single kosmic Avatar of Reality. This is a revelation that has been known by individual human beings, and occasionally by groups, since the moment our reptilian brain stems, limbic systems and neocortices first allowed us to sense the Ultimate Ground and Source of all.. More and more now, though, the pockets in which this realization is arising are growing sufficiently large that its truth can begin to be collectively *enacted*. The question then arises, "Once we know ourselves collectively as the radiant emanation of Awakened Awareness, the Boundless Immutable Principle that is arising as the entire kosmos, what happens then?"

Then, through the whole of our lives, internal and external, individual and collective, we build temples to that realization.

Chapter 23
The New Temples of Eden
Jon

What might humanity and Earth look like if and when such a kosmocentric culture and civilization of Reality has been born here? What might we see when human life, the Earth and kosmos, and the sphere of the Utterly Sacred, are recognized as one? In this penultimate chapter of the book, we explore a vision of the temples that are to come. We imagine these as the spiritual centers of the coming culture in which the radical Truth of the One Reality and the continuing waterfall of evolutionary truths concerning the kosmic significance and community in which humanity is participating, are mined, explored, and transmitted throughout the body of Earth and out into the kosmos. We do so by first considering some of the teaching given in the Trans-Himalayan movement concerning the new mystery schools, or temple communities of the future, that are prophesized to come. We then envision how these temples might look in the context of what we have explored in this book already. This is done structurally, in terms of how some of the ancient wisdom-technology demonstrated in the ancient temples might be re-integrated, synthesized and updated. I also then consider the question functionally, in terms of the role that these coming temples of humanity, Earth, and the kosmos are envisioned to play within the planetary body.

The Return of the Mystery Schools

In the Trans-Himalayan teaching, it is understood that should we navigate our present collective crises successfully, as more and more human beings awaken radically and evolutionarily, and world- to kosmocentric structures of consciousness, behavior, culture, and civilization come online, there will come the rebirth of the ancient mystery schools and temples, but in a new, updated form. As we see it, these will be the temples of the One Planetary Lineage of Reality and the Great Human Tradition. They will be temples to the one humanity as a cosmic bodhisattva; to the sacred community of kosmic civilizations in which we may increasingly participate; to the Earth and its Purpose as an inseminator of the kosmos with Life; and of the radical Truth of Awakened Awareness that saturates every single particle of the All.

According to the multidimensional vision of the Earth we have explored herein, the emergence of these mystery schools, or new temple-communities, relates to the externalization on the physical plane of the ashrams of Hierarchy described in Chapter 11. From this perspective, we might envision each temple as expressive of one of the seven ray energies. Or, they might be expressing various combinations of those energies, and acting within the planetary aura as transmitting stations for them from those constellations that transmit them kosmically, through Shamballa and Hierarchy, and into the Earth-sphere. They would thus each allow the emergence of the physical body of the master incarnating through that ashram, as described in Chapter 11, in which those human beings composing the mystery school temple community can participate. They would act as community spaces for profound training in which human beings could participate in the mysteries of radical and evolutionary awakening in the kosmos as an awakened WE. From the Integral perspective we have explored in this book, these temples involve the full decent into physical matter of each of the LR energy fields of the planetary cultures of Hierarchy, and eventually Shamballa, operating on subtle planes. In the language of the Trans-Himalayan teachings, this is the Externalisation of the Hierarchy and the Reappearance of Christ.

In Djwhal Khul's teachings, these new temples are described as mystery schools, with each being affiliated with one of the global branches of the planetary Hierarchy. Recall here the point made in Chapter 2 that the Himalayan Branch of Hierarchy, from which the Trans-Himalayan teachings have come, is understood to be but one of a number of branches, others likely being affiliated with such global regions as South America, Africa, Southern India, North America, Australasia and Europe. Djwhal Khul teaches that all of the coming mystery schools will have their shared root in the One Fundamental School of Wakefulness and planetary Purpose that is Shamballa. It is additionally described that the masters of Hierarchy envision the emergence of both *preparatory* and *advanced* mystery schools, with the differentiation being based on the level of initiation beings have passed through. Specifically, the first two initiations would be passed through in the preparatory temple schools, whilst the third initiation onwards would be passed through in the advanced temple schools. Here is Djwhal Khul:

> In the preparatory schools soul education takes place, and the result of the work of these schools will be the empowerment of souls into full expression in the three worlds [the mental, emotional and etheric-physical planes]. They will join the conscious creators, the educators, the leaders of their generation and help inaugurate those projects within the outer world which will form the seeds of the coming civilization.
>
> In the advanced schools which will come later, the work will be more removed from the outer plane. This should not be interpreted as being ineffective in the outer world however. On the contrary, the spiritual impact of those working within these schools will be significant. This is because their students have already initiated their own projects in the three worlds in this life (and in some rare cases, in earlier ones), and they will be consciously energizing and empowering those projects through their meditative work. Of course they will be involved in many other meditative activities as a part of their training, but each will have a very specific project that they have created and now continue to serve to inspire, through direct telepathic contact with those who are currently in charge of its functioning in the outer world. And of course these initiates are inspired in turn by those more advanced than they, who laid the seeds for their own self-initiated endeavours just as I have laid out the seeds for the School with which you are engaged, and in turn am inspired. Such is the conscious Hierarchy of the Aquarian Age.
>
> Think this all through. At the Third Degree an initiate creates his outer 'masterpiece' which is the summation of his soul's experience over many lives in the three worlds, and is self-initiated in alignment with ashramic intent. Looking around the world many of these can be seen. They are the humanitarian organisations, the religious movements, the modern businesses, the soul-inspired literature and art. Each soul-inspired project has at its core at least a centre of solar fire and in some cases a 'jewel' of monadic fire. This fire is 'fanned' from the inner worlds, stimulating the ongoing life and inspiration for the outer work via the minds and hearts of those who are in outer 'control'. When those in control are conscious of their

inner inspiration much more can be achieved, and we
will begin to see this success demonstrated as more and
more initiates move into positions of outer authority.[207]

A crucial piece to how these mystery schools, or coming temples,
are understood in the Trans-Himalayan teaching is that their focus,
particularly in the advanced schools, will be on radical awakening
to Absolute Reality and evolutionary awakening into the monadic
altitudes of identification, relationship, plane access and intelligence.
What this means is that in these schools it will be the stabilization of
radical Enlightenment and the electric revelation of monadic Purpose,
Will, Power, and Life at the heart of our entire planetary process that
will be the focus; the full embodiment of radiant divinity, through the
whole personal nature—mind, feeling, and body on our sacred Eden
planet; as well as their cultivation of multidimensional relationship,
plane access, and intelligence unfoldment on extraordinary levels. Such
temples will be fields in which communities of initiates and masters
blossom, revealing the planetary Purpose of the Earth itself through
the heart and body of their awakened We and then transmitting that
realization into the rest of the Earth, and out into the kosmos. In the
words of Djwhal Khul:

> The work of the modern day Mystery school is to plant
> the seeds of [monadic] Will. These seeds will flower in
> the future as the widespread realisation of planetary
> soul purpose: The collective comprehension and
> implementation of the multi-incarnational purpose of
> the Earth as a conscious entity in cosmos.[208]

The seeds of such mystery schools, and Temples of Humanity,
can definitely be seen today. Damanhur in Italy is one example, and
Auroville in India is another. There is so much more to come, though.
In order to invite in more of that which is being envisioned and held as
information on subtle planes, I will now explore a speculative vision of
how these temples may look, both structurally and functionally.

---

207. Lyon, 2005, p. 51-52.

208. Lyon, 2003a, p. 127.

The Structure of the Coming Temples: Honoring the Ancients

If we envision now what such temple-schools might look like structurally, we might first see them as honoring the wisdom embodied in every great temple of the Earth that has gone before, synthesizing and integrating that wisdom, and expressing it in an updated form. To understand what this means in a little more detail, it is first important to honor that the existence of temples in which the Great Mysteries of Life and Reality have been explored and celebrated by the initiated goes back into ancient times. Examples include the pyramid-temples of Egypt and South America; the temples of Ancient Greece, Mesopotamia, and India; and the mountain retreats of Tibet, Syria, and China, among so many others. In each of these places, human beings were initiated into communities of radical and/or evolutionary awakening in which the deepest enigmas of our being could be explored and lived.

A shared characteristic of the ancient mystery schools, however, as was pointed out in the previous chapter, is that owing to the collective radical and evolutionary center of gravity for humanity as a whole, their mysteries largely remained secret and unknown to the majority. Indeed, with the rise of religions such as Christianity and Islam, owing to the strongly ethnocentric worldviews that have been so prevalent in those traditions, the mystery schools that *were* known were often persecuted and eventually driven underground in order to survive.

The most potent temples are spaces in which both humanity and the Earth are mutually empowered through the coming into relationship of the deepest human realizations with sacred power spots of the Earth's body. In this connection, the sophisticated Earth and energy technologies that the ancient builders worked with in constructing their temples were driven underground, and we are only beginning to understand them again now. There is a profound wisdom involved in the art of temple construction, one that entails an understanding of the power already present in the geosphere and biosphere of the Earth. As John Burke and Kaj Halberg describe in their book, *Seed of Knowledge, Stone of Plenty*, it seems that the ancient temple builders, across cultures, built their sacred structures on land spaces with particular geological properties. These often incorporated the abundant presence of materials like limestone, or granite, whose high level electro-conductivity produced intense effects on the Earth's geomagnetic field. Here are the authors speaking about the pyramids of the Egyptian plateau:

> All pyramids were built with an electrically conducting core of coarse local limestone with a high content of

339

magnesium. These core blocks are what we see today, the sides being anything but smooth. This is because, in the thirteenth century Moslems tore off the outer casing and used it for building materials as the start of an ultimately abandoned effort to destroy these pagan structures.

In only a few spots can we still see the beautiful, snow white, fine Tura limestone that made up the outer casing. This stone was finer grained than the stone employed in the core, and polished to a sheen. It came from the East Bank of the Nile, and had to be ferried across. Most of the giant pyramids were ultimately encased in Tura limestone. We find this intriguing because, in contrast to the core limestone, Tura limestone contains only traces of magnesium and is therefore a poor conductor of electricity. In fact, it acts as an *insulator.*

What we have then, is a massive pyramid of electrically conductive limestone blocks, cloaked in an insulating outer layer. This combination would prevent any of the electrical charge in the core from leaking off into the air. The Tura casing blocks are one of the greatest engineering wonders of the ancient world. They were polished on all sides to within $1/100^{th}$ if an inch accuracy, fitting together so snugly that you cannot get a razor blade between them, five thousand years later. The labor involved in this creation was extraordinary. The builders either had an incredible fetish for precision, or this precision was functionally important. With this tight insulating cover, the only place the electrical charge could leak out would be through the benben, [the cap stone] at the apex of the pyramid. All negative charge, spread throughout its entire base would be concentrated in one pointed capstone, possibly of pure iron....

The geographical sitting of the ancient pyramids was hardly a coincidence. All were built atop a limestone plateau on the west side of the valley. They all had contact with the limestone bedrock. Underground, the rock strata tilts downward toward the river, forming the edge of the Nile Valley aquifer. As we know, water moving up and down an aquifer creates electrical currents.[209]

The discussion of the material properties of the pyramid complex at Giza by Burke and Halberg above indicates that the ancient

---

209. Burke & Halberg, 2005, p. 159-161.

Egyptian temple builders may have known very well what they were doing by situating the Pyramids where they did. By building the pyramid complex on the Giza plateau, and by using the combination of materials they did, they would be able to create a virtual *power station.* By building in the specific locations they chose, employing the methods for this they did, and using the specific materials they worked with, the electrical charge created by the underground aquifer below the plateau, as well as the electrically charge concentrated in the magnesium of the limestone blocks, would be potently concentrated within the pyramid, and funneled upwards so as to flow out through its apex. These researchers found similarly fascinating examples of the ancient temple technologies used at such sites as the pre-Colombian Maya pyramid of Tikal, Tiwanaku of Titicaca in what is now Bolivia, the henges of Southern England, and the mounds of North America. Each of these shows sophisticated knowledge on how to harness and concentrate the electromagnetic energy of nature.

All of this is deeply pertinent to the role of such temples as places in which the radical and evolutionary awakening of human beings could be greatly empowered. According to the research of neuroscientist Dr. Michael Persinger, fluctuations in electromagnetic fields surrounding human beings can have dramatic effects on their consciousness. Indeed, Persinger has suggested that many instances of apparent psychic activity, telepathy, or the spontaneous occurrence of hallucinations can be explained according to the effects of anomalous geomagnetic activity of the Earth on the brain. It is likely that the neurobiological effects of such concentrated Earth energies on human consciousness would have extended across both the radical and evolutionary vectors and all lines. This would mean that when the ancient temples were in their former maintained state, they would have been spaces in which the possibility of radical awakening and evolutionary shifts in identification, the opening up of multidimensional relationship, plane access, and the unfoldment of the intelligences of the subtler planes were potently *amplified.* Additionally, these places may therefore also have held powerful fields for the health, evolution, and awakening of all lives in their surrounding areas.

In this respect we might envision the coming temples of Earth, once it has been revealed as Eden and is being honored by a kosmocentric culture and civilization of Reality, as integrating the wisdom of the ancients in how the structure of the coming temples of the future emerge. More specifically, we could envision that the new structures would honor the truths embodied in both the temples of ascent, such as the upward pointing pyramids and temples of ancient Egypt, South America, India and the mountain sky-retreats of Tibet. We could envision that they would also incorporate those of the temples of descent, such as the ancient Celtic and American Indian sacred spaces, which are built deep into the Earth. In this connection, Djwhal Khul and Bruce Lyon suggest that the coming temples may be constructed in the form of a giant octahedron, which both penetrates into the sky and into the core of the Earth:

> The first phase of the unfoldment of the modern Mystery traditions was anchored under impulse from the Sirian lodge and has been held at the core of many systems of civilisation; in particular the Vedas, the Egyptian mysteries and the systems of South America. The primary focus of the Mystery traditions has been the initiation into the soul, which is why the sun has played such an important symbolic role. These traditions and initiations are fast becoming exoteric. The current dispensation and the return of the modern mystery schools focuses upon the Will, the monad, and will result in what has been termed the Temple of Power manifesting on the physical-etheric plane. The pyramid is a geometrical symbol for the first stage of the mysteries anchoring. The octahedron (outside the double tetrahedron) is the symbol for the second stage. It represents the coming of cosmic fire to Earth.[210]

One of the profound glories of the ancient Egyptian temples is that in their wondrous mathematical and geometrical precision they did not just mirror the harmony of the kosmos, but they actually *participated in and contributed to it*. In keeping with this, we might envision these coming temples to also engage this approach of 'kosmic feng shui'. As such they would employ sacred geometry so that their structure is naturally resonant with the numerical and musical patterns that Ultimate Reality Itself displays energetically in its involutionary path so as to arise as every form. Biomimicry would be another approach to structure that we envision would play a key role, where the wisdom of nature is also honored and harnessed. Such temple buildings would embody physical yogic asanas of the Earth whose posture and form are so perfectly balanced that they open harmonic pathways into the timeless and evolutionary realities of the kosmos.

---

210. Lyon, 2010, p. 214-215.

The Function of the Coming Temples: A Vision for the Future that is Possible

Here we will consider how the coming Temples of Eden might look functionally, and with reference to both the radical and evolutionary vectors and lines.

Firstly, such temples would be spaces in which the Truth, entailed in radical awakening to the infinite Conscious Light of Absolute Reality— Awakened Awareness, the One Life—formed the basic essence of the entire Temple.

In terms of evolutionary awakening, we could first envision them as sacred spaces in which both the altitude of monadic identification held by the community of initiates residing there, and their partnership with the land in working with the powerful materials of the sacred Earth, as was described earlier, would be deeply important. These would be such that the very heart-mind and Presence of the planetary Logos of the Earth, and the Purpose envisioned by that great being, became so anchored and amplified in those spaces that its Presence would be known by all as the source and basis of their own monadic identity.

In connection with the relationship line, we would envision them as spaces in which collaborative relationship and sacred friendship in planetary service was honored with the multidimensional evolutionary communities residing within the aura of the Earth. We foresee such temples as spaces of sacred regard for the other kosmic humanity's and civilizations with whom we serve galactically, as part of a kosmic Federation of Light. As such, in these temple spaces would be preserved and honored the ancient stories of evolutionary becoming and radical awakening of all the communities, cultures, and civilizations, both Earthly and kosmic, with whom we stand as a family of Light, Life, and sacred Form.

Along the line of plane/energy access, we can envision these temples as lightning rods for kosmic energies into the Earth sphere, into the core of the Earth. Such temples might be structurally aligned with the constellations transmitting the specific energies that they seek to transmit. This was the case with the constellations of Orion, Sirius and the Great Bear with the Great Pyramid in Egypt, and that might today relate to alignment with such constellations as these again, or the galactic center.

Along the intelligence line, it is envisioned that within such temples might be spaces in which light-technologies are used to teach some of the most profound mysteries of Reality, humanity, and the kosmos. This will only be possible when humanity has begun to unfold the intelligences of the soul planes. It would allow the temples to act as educational spaces in which, perhaps, the truths of the kosmos and each kosmic civilisation are demonstrated in holographic forms of light and kosmic maps. These might demonstrate the star systems where each originate, as well as the universal wisdom transmissions that Reality itself is pouring forth in every single second from the hearts of suns, galaxies, planets, human beings and every life form, right down to and beyond the atom.

Of course, that which is presented here should be taken lightly. It is but an open speculation of what one day may be possible, very likely long into the future, should humanity emerge from its current challenges to unite as a single family in stewardship of this planet. Regardless of the details explored here, should that happen we will no doubt come to see a single planetary future of the sacred, embodied in temples of awakening, transformation, evolution and communion around the globe. In this connection, we recall once again the words of the Master Morya quoted at the beginning of Section 1:

> You and We—here together in spirit.
> One Temple for all—for all, One God.
> Manifold worlds dwell in the Abode of the Almighty,
> And the Holy Spirit soars throughout.
> The Renovation of the World will come—
> the prophecies will be fulfilled.
> People will arise and build a New Temple.

Chapter 24
Conclusion: Earth is Eden
Jon and Dustin

Having now journeyed this far with us, in this concluding chapter of the book let us revisit two of the core intentions that we stated were at its heart. In the welcome section we noted that "the primary intention behind this book is to contribute to the revelation of the kosmic Purpose of the Earth, and the emergence of an awakened, kosmocentric culture and civilization of Reality on Earth". From the root vantage point of Primordially Awakened Awareness and an evolutionary altitude of monadic identification, Earth is revealed as Eden. As the path of evolutionary becoming unfolds within the Infinite Matrix of Absolute Presence, our shimmering, blue-green planet is continuously and forever evolving into greater and greater expressions of perfection.

In the welcome section, we also noted that,

> ... in today's world, the One Universal Revelation of Reality has been imprisoned by ethnocentric and at times pathological expressions around the globe. This ethnocentric imprisonment, and the tribalism that results, is one of the single greatest contributing factors to world conflict and suffering. As such, this book is a contribution towards the emergence of a universal spirituality on Earth. We envision such a universal spirituality not as a homogenized mass in which the differentiations between points of wisdom in the spiritual traditions are lost. Rather, we see it as one that can see, honor, and even optimize those points of differentiation from a collective space that recognizes that different cultures have pioneered different forms of expertise over the ages

when it comes to awakening, transformation, and the exploration of reality.

In Chapter 1, we included a diagram that points to a trajectory by which each of the Earth's spiritual traditions can provide pathways into the trans-lineage space of a universal spirituality.

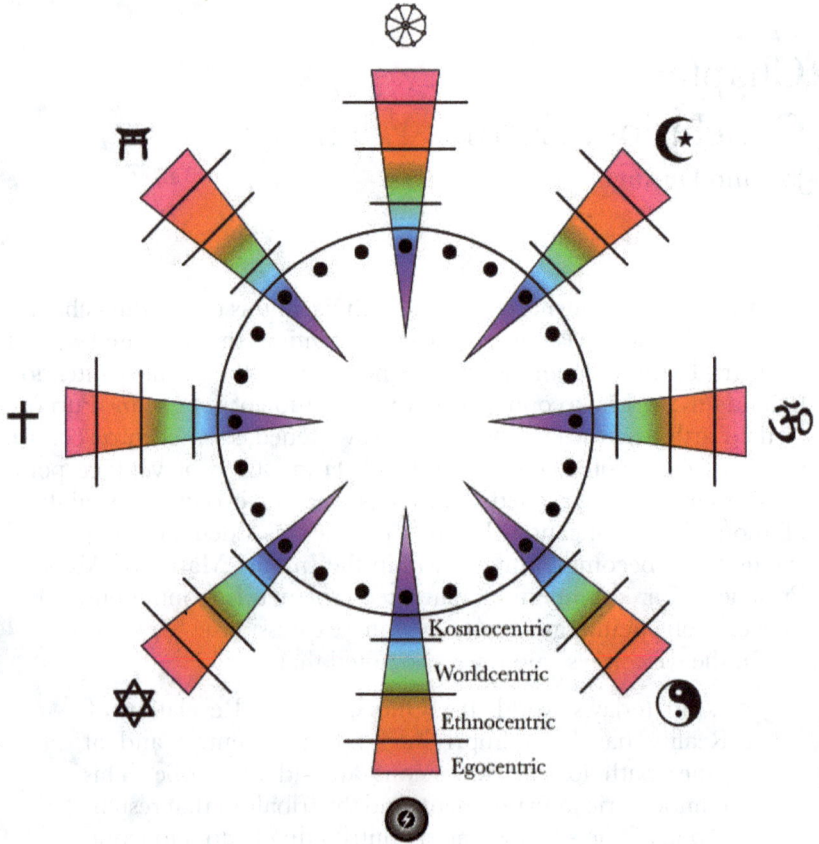

Figure 1. Graphic showing the pathway for each of the spiritual traditions from egocentric to ethnocentric to worlcentric to kosmocentric expressions. In the latter, they each offer doorways into a universal spirituality. The points that do not stand within lineage streams represent those who have entered into a universal spirituality without particular connection to a tradition. The lightning symbol at the bottom represents the Trans-Himalayan pathway. The Shamballa School symbol is used to represent it as there is no single symbol for the whole tradition.

As Ken Wilber and Dustin have pointed to in previous books,[211] spiritual traditions occupy a unique space in culture. They are the only domain able to act as conveyor belts for human beings not only to access states of realization, all the way through to the Primordial State of Awakened Awareness, but that are also able to serve as one of the primary facilitators of growth through evolutionary structures. This points to their capacity to act as stewards and guides of human beings into their most awakened, integrated, and evolved expressions in a way that no other domain of culture and society caters for.

In order for traditions to play this role, they must have unfolded expressions of themselves through all state/phases of Reality and through all evolutionary structure-stages available to humanity at the present time. Specifically, this requires them not only to provide pathways to radically awaken as the One Infinite Presence via state/phases, but also to unfold not just traditional, modern, and post-modern expressions, but integral, holistic, world-centric into kosmocentric expressions as well. This can only be done as representatives and practitioners within the traditions pioneer these new territories themselves and co-create, in alignment with the particular texture and feel of the flame of liberation at the heart of their tradition, what the higher structure expressions of it look like.

This has been a developing project through all the traditions for some time now. If we trace it through the structure-stages unfolding through the mental plane that we discussed in Chapter 13, we can see that at the traditional structure-stage (amber), ethnocentric adherents to a particular path normally see their own way as absolute, with other paths simply being wrong, and therefore not of much interest. In the modern structure-stage (orange), we start to see this change. Here, as worldcentric consciousness emerges, the study of comparative religion and comparative spirituality begins to emerge. It is at this stage that scholars analyze how the different pathways compare to each other, but often remain at a distance from other traditions and relate to them in an objective third-person way. In the pluralistic structure-stage (green) this changes again, and through inter-faith work representatives of different pathways come into relationship with each other. These relationships include the comparative, intellectual component of the exchange, but they are not limited to it. They also include a much greater degree of intimacy in the form of 2nd person contact with each other. As these relationships between representatives of the different traditions develop, more and more is learned about the great diversity inherent within the various streams of spiritual traditions. In some forms of

---

211. See Ken Wilber's *Integral Spirituality* and Dustin DiPerna's *Evolution's Ally* *(Volume 2 of this series)*.

pluralism, representatives begin to recognize the fact that each of the various traditions provides a window into a particular aspect of the One Reality.

This is a crucial development, one that holds incredible promise to contribute to the end of the conflict, strife and suffering that pervades the relationships between the different traditions that has been so prevalent through history. This is the point where, as we have said previously, each spiritual tradition is recognized as a branch of the Great Human Tradition. However, owing to the tendency of the pluralistic structure-stage to honor the diversity of different perspectives without simultaneously being able to differentiate them according to their degrees of depth, the path of growth can still be taken further.

This is what happens at the integral structure-stage (turquoise). Here, often with the help of such meta-frameworks as Integral Theory, and as is happening more and more today in a manner that represents one of the seeming leading edges of human evolution, representatives of the different lineages and traditions have begun to go beyond merely being in relationship with each other. As such, these representatives are now allowing all traditions to enter into their own direct 1st-person experience. They have come to recognise the different traditions as streams of one trans-lineage river. However, instead of seeing them as just different but essentially equivalent windows on the same reality there is the recognition that these different approaches have cultivated extraordinary insights in relation to different aspects of life, from spiritual awakening to cultivating a deep relationship with the divine, to psychological wholeness and healing, to what it means to open up access to multiple realms, and participating in the evolution of culture and society. With this, there is the additional recognition that when we put them together, a new, truly global and integrative approach to human growth and awakening emerges that has never been seen before. In the diagram above, this takes us to the line that borders those sections of each of the traditions, represented by the elongated triangles, which exist outside and within the central circle.

Transition into the central circle involves a movement that is only just beginning to be engaged by certain pioneering representatives of the different spiritual traditions. At an indigo structure-stage of development representatives remove their robes and nakedly step into the River of Universal Light. At this stage, because each representative is first and foremost a human being, all the great lineages of human wisdom are seen as mapping possibilities of one's very own 1st-person potential.

Pioneers of this process are beginning to explore what it means not just to integrate the offerings of the different streams side by side with each other, but to bathe in this one river of naked Reality-communion and identification with the one humanity that is their source. This new approach includes the synthesized essences of all the lineages but in a way that is no longer defined by the cultural history and story in which they have lived. As the perspectives available in the transpersonal structures of evolutionary development come fully online, they are kosmocentric.

This is the path that is available to the spiritual traditions and pathways so that they might provide the most stable conveyor belts for human beings to tread the Way into Truth, from egocentric to ethnocentric to worldcentric to kosmocentric. In the words of Djwhal Khul, "A spiritual tradition is like a river that leads to the ocean. Once the destination is reached then the banks of the river disappear."[212]

As was pointed to in Chapter 2, however, unlike the rest of the traditions that Djwhal Khul is pointing to, the Trans-Himalayan teachings were never intended to become another tradition or pathway. They were never intended to indefinitely remain as another of the rivers leading back to the ocean. Rather, like a flash of lightning hitting the surface of the ocean so that its blast of electric current pours back up the rivers that pour into it, they were designed to inform the universal space, and the lineages of the traditions themselves, with a transmission of our planetary Purpose, a kosmocentric worldview, and the spirit of multidimensional planetary cooperation. The pathway that the Trans-Himalayan teachings have provided is simply a temporary one to support its practitioners' penetration into radical wakefulness and contact with the power of monadic identification. It was not designed to provide a conveyor belt path that would remain around for centuries or millennia but rather, on the basis of the voltage of current generated from its transmission of monadic Purpose, to consume itself in flames and thus inform the trans-lineages space of a universal spirituality and the planetary field as a whole with its essence. This occurs through its practitioners' increasingly stable radical awakening to Source and contact with the electric fire of monadic identification, expressed in consciousness, daily life, relationships and service on every level (through all Four Quadrants).

The teachings of a tradition reside as a galaxy of thought-structures on the mental plane, and technically we could say that it is an increasingly stable Reality realization and capacity to handle monadic current that actually transfigures and burns through the subtle body of a tradition as it exists on the mental plane. As students of the Trans-Himalayan teaching radically awaken and draw on the power of monadic identification in their individual and collective processes, a beam of fire burns through that galaxy of thought-structures on the

---

212. Lyon, 2010, p. 319.

mental plane, where its core concepts exist. This allows the atmic will-force and buddhic wisdom-essences at the core of those thought-forms to be released out into the collective field of humanity, no longer held within the mental thought-forms in which they have been clothed since the inception of the teachings. Through that release, the wisdom-truths that were anchored in the mental plane in conceptual frameworks become able to be contacted and recognized intuitively by evolving humanity at large.

This is the process that the Trans-Himalayan teachings were initiated to pioneer. As such, we can see that they could be considered as just a house that was built to provide a structure within which the truths of the Hierarchical teaching could be made available to humanity, and held within reach of human functioning for a specific period of time. That period would be long enough for those working with the teachings to radically awaken and learn to be able to work with monadic Life, so as to begin to burn through the thought-forms in which its transmission is housed. And simultaneously, it would be long enough for humanity as a whole to evolve enough to be able to contact the wisdom-essences that would then be released onto the buddhic plane via intuitive mind after the burning of the mental sheaths in which they had been anchored. I would suggest that the Trans-Himalayan teachings have thus been a temporary mental plane anchor employed by the liberated masters and guides of Hierarchy for these truths until humanity could grow enough to be able to contact their buddhic wisdom-essences to claim them as its own rightful inheritance. For those affiliated with any tradition or none, this all makes plain the power of working with monadic Life from radical awakening. As we do so, we burn through the webs separating humanity from Truth.

Thus, one of the central injunctions of this book is to call the students of the Trans-Himalayan teachings who understand their potential to powerfully inform the planetary field, to a new and ever-renewing depth of commitment in their practice, life, and service. That means to investigate the Source and True Nature of their awareness and find it to be God—the Ultimate Self-Condition of Reality that is emanating as the entire kosmos. From that innate freedom and completeness, it means progressively opening up the multidimensional altitudes of our identity, relationship, plane access and intelligence. It means to live a life of such soulful dignity, in consciousness, behavior, culture and outer service, that the lightning transmission of monadic identification is evoked to empower the raw heart of our innate divinity, and the collective process of planetary service to which the Trans-Himalayan teachings are committed. Furthermore, it means doing this with purest integrity, utterly committed to the healing and integration of any areas of shadow in our systems that distort the Clear Light's radiant expression through our multidimensional relative being.

To the degree that those committed to serving the purpose of the Trans-Himalayan teachings engage this process, we will see them demonstrating radical awakening and monadic empowerment in the unique colors and form of their individual lives. We will see the gradual emergence of a culture of radically awakened monadic fellowship. This would be a community of mature and multidimensionally integrated men and women through whose Reality-Hearts the lightning of our planetary Purpose is pouring. Such a group would be able to act increasingly on the planetary field like a black hole warping time and space in conformity with its purpose. This is the emergence of the trans-lineage group initiate that will steward the final transmission of the teachings that will come around 2025, and that will mark the end of the Trans-Himalayan movement.

As this occurs, the incense of the Trans-Himalayan teachings will be burned on the altar of a universal spirituality; the wisdom-essences they have anchored on the mental plane will be released so as to enrich the trans-lineage space and to be contacted intuitively on the buddhic plane by increasing numbers of human beings. It is this living process that is intended to inform the universal space of trans-lineage communion with a transmission of our planetary Purpose, and to burn truth clean of mental plane concepts sufficiently for more and more human beings to contact it without passing through the conveyor belt of a tradition. Over the course of time, through the kosmocentric updating of the traditions, as well as the awakening and evolution of the human beings within them and of no tradition, a universal spirituality may come forth.

The planetary crisis we currently face is deep. There are many variables involved. The spiritual pathways of the Earth no doubt shoulder a profound responsibility to support humanity at large to move beyond tribal conflict and discover our birthright of awakened sovereignty and planetary friendship. The stakes are high, and the consequent opportunities are great. Should we together navigate the present crises we face so as to heal and integrate our world, as the Trans-Himalayan prophecy suggests, we will have the opportunity to enter an age where humanity will no longer need any intermediary between itself, recognized Ultimate Reality, and the revelation of kosmic Purpose through all of life. Then, should we together make it that far, we may see the emergence of a radically awakened kosmocentric culture and civilisation of the One on Earth, and our role in the galaxy as a kosmic bodhisattva will have begun.

References

Adi Da Samraj (1978). *The Enlightenment of the Whole Body*. CA: Dawn Horse Press.

Alien Habitable Planets May Exist Billions of Years Older Than Earth --Harvard Center for Astrophysics. (2012, June 13). *The Daily Galaxy*. Retrieved from http://www.dailygalaxy.com/my_weblog/2012/06/alien-habitable-planets-may-exist-billions-older-than-earth-harvard-center-for-astrophysics.html

Allen, J. P. (2005). *The Ancient Egyptian Pyramid Texts*. Atlanta: Society of Biblical Literature.

Are Humans the Limit of Evolutionary Complexity? (2009, October 10). *The Daily Galaxy*. Retrieved from http://www.dailygalaxy.com/my_weblog/2009/10/are-humans-the-limit-of-evolutionary-complexity.html

Arsenic loving bacteria may help in hunt for alien life. (2010, December 2). *BBC News*. Retrieved from http://www.bbc.com/news/science-environment-11886943

Aurobindo, G. (1996). *The Synthesis of Yoga*. Twin Lakes, WI: Lotus Light Publications.

Aurobindo, G. (1970). *Letters on Yoga Volume 1*. Pondicherry: Sri Aurobindo Birth Centenary Library.

Aurobindo, G. (1989). *The Psychic Being: The Soul in Evolution*. Pondicherry: Sri Aurobindo Ashram Trust.

Bailey, A. A. (1925). *A Treatise on Cosmic Fire*. New York, NY: Lucis Publishing Company.

Bailey, A. A. (1930). *The Soul and its Mechanism*. New York, NY: Lucis Publishing Company.

Bailey, A. A. (1934). *A Treatise on White Magic*. New York, NY: Lucis Publishing Company.

Bailey, A. A. (1942). *Esoteric Psychology Volume 2*. New York, NY: Lucis Publishing Company.

Bailey, A. A. (1950). *Telepathy and the Etheric Vehicle*. New York, NY: Lucis Publishing Company.

Bailey, A. A. (1955). *Discipleship in the New Age Volume 2*. New York, NY: Lucis Publishing Company.

Bailey, A. A. (1957). *Externalisation of the Hierarchy*. New York, NY: Lucis Publishing Company.

Bailey, A. A. (1960). *The Rays and Initiations*. New York, NY: Lucis Publishing Company.

Bhattacharya, K. (2015). *The Atman-Brahman in Ancient Buddhism*. Cotopaxi, CO: Canon Publications.

Blake, H. (2010, February 22). Royal astronomer: 'Aliens may be staring us in the face'. *The Daily Telegraph*. Retrieved from http://www.telegraph.co.uk/science/space/7289507/Royal-astronomer-Aliens-may-be-staring-us-in-the-face.html

Blavatsky, H. P. (1928). *The Secret Doctrine Volume 1*. London: Theosophical Publishing House.

Brown, Daniel. (1981). *Mahamudra Meditation Stages and Contemporary Cognitive Psychology: A Study in Comparative Psychological Hermeneutics*. Dissertation: University of Chicago.

Burke, J., & Halberg, K. (2005). *Seed of Knowledge, Stone of Plenty*. San Francisco: Council Oak Books.

Campos, D. J., Grob, C., & Roman, A. (2011). *The Shaman and Ayahuasca: Journeys to Sacred Realms*. CA: Divine Arts.

Cedercrans, L. (2004). *Ashramic Projections*. Whittier, CA: Wisdom Impressions.

References

Chau, T. (1999). *The Literature of the Personalists (Pudgalavādins) of Early Buddhism*. New Delhi: Motilal Banarsidass Publishers.

Darling. D. (2001). *Life Everywhere: The Maverick Science of Astrobiology.* New York, NY: Basic Books.

Diamond Star Thrills Astronomers. (2004, February 16). *BBC News*. Retrieved from http://news.bbc.co.uk/2/hi/3492919.stm

Did a Supernova Shockwave Create Our Solar System? New Finding Says "Yes". (2012, August 4). *The Daily Galaxy*. Retrieved from http://www.dailygalaxy.com/my_weblog/2012/08/did-a-supernova-shockwave-create-our-solar-system-new-findings-say-yes.html

DiPerna, D. (2014). *Streams of Wisdom*. San Francisco, CA: Integral Publishing House

DiPerna, D. & Augustine, H. B. (2014). *The Coming Waves*. San Francisco, CA: Integral Publishing House

DiPerna, D. (2015). *Evolution's Ally*. San Francisco, CA: Integral Publishing House

Emanuel, M. S. (2006). *Divine Design*. Southfield, MI: Targum Press.

"Evidence of Alien Life Expected Within 21st Century" --Leading Astrophysicist. (2012, June 16). *The Daily Galaxy*. Retrieved from http://www.dailygalaxy.com/my_weblog/2012/07/evidence-of-alien-life-expected-within-21st-century-leading-astrophysicist.html

Farb, N. A., Segal, Z. V., Mayberg, H., Bean, J., McKeon, D., Fatima, Z., et al. (2007). *Attending to the present: Mindfulness meditation reveals distinct neural modes of self-reference*. SCAN, 2, 313-322. doi:10.1093/scan/nsm030

Gamma Rays. (n.d.). *The Ozone Hole*. Retrieved from http://www.theozonehole.com/gamma.htm

Johnson, K. Paul. (1998). *Edgar Cayce in Context: The Readings: Truth and Fiction*. Albany, New York. SUNY Press.

Krishnananda (2008). *The Mandukya Upanishad: An Exposition*. Rishikesh: The Divine Life Society.

Lyon B. P. (2003a), *Mercury*. Wellington, NZ: White Stone Publishing.

Lyon, B. P. (2003b), *Agni: Way of Fire*. Wellington, NZ: White Stone Publishing.

Lyon, B. P. (2005). *Group Initiation*. Wellington, NZ: White Stone Publishing.

Lyon, B. P. (2007). *Working with the Will*. Wellington, NZ: White Stone Publishing.

Lyon, B. P. (2010). *Occult Cosmology*. Wellington, NZ: White Stone Publishing.

Mantak Chia. (n.d.). *Darkness Technology: Darkness techniques for enlightenment*. Chiang Mai: Thailand. Universal Tao Publications. Retrieved from http://www.universal-tao.com/dark_room/DarkRoom.pdf

Newton, M. (1994). *Journey of Souls*. St. Paul, MN: Llewellyn Publications.

Newton, M. (2001). *Destiny of Souls*. St. Paul, MN: Llewellyn Publications.

Norfleet, P. (n. d.). *Consciousness Concepts of Gerald Edelman, Academic Studies of Human Consciousness*. Retrieved from http://consciousness2007.tripod.com/gerald_edelman.htm/.

Reynolds, J. M. (1996). *The Golden Letters*. Ithaca, NY: Snow Lion Publications.

Ridley Scott's "Prometheus" Suggests DNA May Be a Constant in the Universe --Richard Dawkins and Other Scientists Agree. (2012, June 10). *The Daily Galaxy*. Retrieved from http://www.dailygalaxy.com/my_weblog/2012/06/-ridley-scotts-prometheus-suggests-dna-may-be-a-constant-in-the-universe-richard-dawkins-and-other-s.html

Robbins, M. D. (2004). *Tapestry of the Gods IV: On the Monad*. Jersey City Heights, NJ: University of the Seven Rays Publishing House.

Roerich, H. I. (1924). *Leaves of Morya's Garden 1*. USA: Agni Yoga Society.

Roerich, H. I. (1930). *Infinity 1*. USA: Agni Yoga Society.

References

Some, P. M. (1995). *Of Water and Spirit: Ritual, Magic and Initiation in the Life of an African Shaman*. London: Penguin.

The Milky Ways' Alien Planets: 160 Million and Counting! (2012, June 23). *The Daily Galaxy*. Retrieved from http://www.dailygalaxy.com/my_weblog/2012/06/alien-milky-way-planets-160-billion-and-counting-weekend-feature.html

Three Jewels Transmissions. (2006). *Shamballa School*. Retrieved from http://www.shamballaschool.com/?page_id=968

Tsoknyi Rinpoche (2003). *Fearless Simplicity: The Dzogchen Approach to Living Freely in a Complex World*. Rangjung Yeshe Publications. Kathmandu, Nepal.

Tulku Urgyen Rinpoche (1999). *As It Is: Volume 1*. Ranjung Yeshe Publications, Kathmandu, Nepal.

Vidal, C. (2010, October). *Black Holes: Attractors for intelligence*. Paper presented at the Kavli Royal Society International Centre, UK. Retrieved from http://arxiv.org/pdf/1104.4362.pdf

Villard, R. (2011, April 30). Super-Civilizations Might Live Off Black Holes. *Discovery News*. Retrieved from http://news.discovery.com/space/super-civilizations-might-live-off-black-holes-110430.htm

Visser, F. (2003). *Ken Wilber: Thought as Passion*. Albany: State University of New York Press.

Wangyal, T. (2002). *Healing with Form, Energy, and Light*. Ithaca, NY: Snow Lion Publications.

Wilber, K. (1977). *The Spectrum of Consciousness*. Wheaton, IL: Quest Books.

Wilber, K., Englar, J., & Brown, D. P. (1986). *Transformations of Consciousness: Conventional and contemplative perspectives on development*. Boston, MA: Shambhala Publishing.

Wilber, K. (1995). *Sex, Ecology, Spirituality: The Spirit of Evolution*. Boston, MA: Shambhala Publications.

Wilber, K. (1998). *The Marriage of Sense and Soul: Integrating Science and Religion*. New York, NY: Random House.

Wilber, K. (2000a). *One Taste: Daily Reflections on Integral Spirituality*. Boston, MA: Shambhala Publications.

Wilber, K. (2000b). *Integral Psychology: Consciousness, Spirit, Psychology, Therapy*. Boston, MA: Shambhala Publications.

Wilber, K. (2003). *An Integral Age at the Leading Edge, Kosmos Trilogy Vol 2, Excerpt A*. Retrieved from http://www.kenwilber.com/ Writings/PDF/ExcerptA_KOSMOS_2003.pdf/.

Wilber, K. (2006). *A Comprehensive Theory of Subtle Energies*. Retrieved from http://wilber.shambhala.com/html/books/Kosmos/ excerptG/part1.cfm/.

Wilber, K. (2007). *Integral Spirituality: A Startling New Role for Religion in the Modern and Postmodern World*. Boston, MA: Integral Books.

World-Leading Physicist Says Extraterrestrials "Could Exist in Forms We Can't Conceive". (2010, September 28). *The Daily Galaxy*. Retrieved from http://www.dailygalaxy. com/my_weblog/2010/09/world-leading-physicist-says-extraterrestrials-could-exist-in-forms-we-cant-conceive-todays-most-pop.html

Zihl, J., D. Cramon, D. V., & Mai, N. (1991). *Disturbance of movement vision after bilateral posterior brain-damage - further evidence and follow-up observations*. Brain, 114: 2235-2252.

www.ingramcontent.com/pod-product-compliance
Lightning Source LLC
Chambersburg PA
CBHW050450270326
41927CB00009B/1686